HANDBOOK OF UNIVARIATE AND MULTIVARIATE DATA ANALYSIS AND INTERPRETATION WITH SPSS

HANDBOOK OF UNIVARIATE AND MULTIVARIATE DATA ANALYSIS AND INTERPRETATION WITH SPSS

Robert Ho

Central Queensland University
Rockhampton, Australia

Chapman & Hall/CRC
Taylor & Francis Group
Boca Raton   London   New York

Published in 2006 by
Chapman & Hall/CRC
Taylor & Francis Group
6000 Broken Sound Parkway NW, Suite 300
Boca Raton, FL 33487-2742

© 2006 by Taylor & Francis Group, LLC
Chapman & Hall/CRC is an imprint of Taylor & Francis Group

No claim to original U.S. Government works
Printed in the United States of America on acid-free paper
10 9 8 7 6

International Standard Book Number-10: 1-58488-602-1 (Hardcover)
International Standard Book Number-13: 978-1-58488-602-0 (Hardcover)
Library of Congress Card Number 2005054817

Library of Congress Cataloging-in-Publication Data

Ho, Robert.
    Handbook of univariate and multivariate data analysis and interpretation with SPSS / Robert Ho.
       p. cm.
    Includes bibliographical references and index.
    ISBN 1-58488-602-1 (alk. paper)
    1. Analysis of variance--Computer programs--Handbooks, manuals, etc. 2. SPSS (Computer
file)--Handbooks, manuals, etc. I. Title.

QA279.H6 2006
519.5'3--dc22
                                      2005054817

Taylor & Francis Group
is the Academic Division of Informa plc.

Visit the Taylor & Francis Web site at
http://www.taylorandfrancis.com

and the CRC Press Web site at
http://www.crcpress.com

Preface

Most statistics textbooks focus on theoretical discussion and mathematical formulas of the concepts presented. The result of this approach is that students often have little understanding of how to *apply* statistical tests to their experimental findings or how to *interpret* their findings. The advent of the SPSS statistical package has gone some way toward alleviating the frustration that many social sciences students feel when confronted with the task of analyzing data from research projects. However, the problems of test selection, execution of a specific statistical test, and the interpretation of results from computer printouts, still remain daunting for most students. The aim of this book is to provide clear guidelines to both the execution of specific statistical tests and the interpretation of the findings by (1) explaining clearly the purpose of specific tests and the research designs for which they are relevant, (2) demonstrating clearly the instructions for executing univariate and multivariate statistical tests, and (3) explaining how obtained results should be interpreted.

The book has been written for use with the SPSS for Windows statistical package. While there are similar books on the market, their instructions for executing statistical tests tend to be limited to an explanation of the Windows method (point-and-click). This book, however, focuses on both the Windows method and the syntax method. The inclusion of SPSS syntax files is based on the following reasons: First, when conducting complex analysis, the ability to write and edit command syntax is advantageous, such as when the researcher wishes to repeat an analysis multiple times with minor variations. Second, the use of syntax files is unsurpassed for data manipulation in complex experimental designs. Finally, from my teaching experience with SPSS, I believe that students have a better "feel" for statistics and experimental designs by writing syntax files to generate the relevant statistics.

I believe that the coverage and presentation format of this book provide a clear guide to students in the selection of statistical tests, the execution of a wide range of univariate and multivariate statistical tests via the Windows and syntax methods, and the interpretation of output results.

Note: The data sets employed in the examples in this book can be accessed and downloaded from the following Website: http://www.crcpress.com/ e_products/downloads/download.asp?cat_no=C6021.

Robert Ho

About the Author

Robert Ho received his Ph.D. from Waikato University, New Zealand, in 1978. He is presently an associate professor in the School of Psychology and Sociology at Central Queensland University, Australia. His teaching and research interests include quantitative methods, health psychology, and social psychology.

Table of Contents

1

Inferential Statistics and Test Selection

1.1 Introduction

The main theme of this book is statistical analysis, which involves both *descriptive statistics* and *inferential statistics*. The major concern of descriptive statistics is to present information in a convenient, usable, and understandable form. For example, once data have been collected, the first things that a researcher would want to do is to calculate their *frequency*, to *graph* them, to calculate the *measures of central tendency* (means, median, and mode), to calculate the *dispersion* of the scores (variances and standard deviations), and to identify *outliers* in the distribution of the scores. These procedures are called descriptive statistics because they are aimed primarily at describing the data. Inferential statistics, on the other hand, is not concerned with just describing the obtained data. Rather, it addresses the problem of making broader generalizations or inferences from sample data to population. This is the more complicated part of statistical analysis, and this chapter will focus on the role that inferential statistics play in statistical analysis.

1.2 Inferential Statistics

As stated earlier, descriptive statistics is used to describe a set of data in terms of its frequency of occurrence, its central tendency, and its dispersion. Although the description of data is important and fundamental to any analysis, it is not sufficient to answer many of the most interesting problems that researchers encounter. Consider an experiment in which a researcher is interested in finding whether a particular drug can improve people's memory. The researcher offers the drug to one group but not to the control group, and then compares the means of the two groups on a memory test. Descriptive statistics will not tell the researcher, for example, whether the difference between a sample mean and a hypothetical population mean, or the difference between two obtained sample means, is small enough to be explained by chance alone or whether it represents a true difference that might be attributable to the effect of the experimental treatment, i.e., the drug. To address these issues, the researcher must move beyond descriptive statistics and into the realm of inferential statistics, and, particularly, on to the statistical procedures that can be employed to arrive at conclusions extending beyond the sample statistics themselves. A basic aim of inferential statistics then is to use the sample scores for *hypothesis testing*.

1.2.1   Hypothesis Testing

Fundamental to the strategy of science is the formulation and testing of hypotheses about populations or the effects of experimental conditions on criterion variables. For example, in an experiment designed to investigate gender differences in IQ, the researcher hypothesizes that first-grade girls have higher IQ scores than first-grade boys. She administers an IQ test to four boys and four girls in a first-grade class. The results showed that the mean IQ score for the girls (110) is higher than the mean IQ score for the boys (103). Based on these findings, is the researcher justified in concluding that her hypothesis is supported? Obviously, the answer is "we do not know." That is, although the results clearly showed that there is a difference between the sample means, with the girls scoring higher on average than the boys, there is the possibility that the observed difference could have been due to the chance variability of intelligence among first graders. In other words, given the variability of intelligence among first graders and the smallness of the sample size, some difference in the means is inevitable as a result of the selection procedures. In order for the researcher to draw valid conclusions about her hypothesis, she needs to employ inferential statistics to test the reliability of the finding of apparent difference in intelligence among first graders. The critical question that must be answered by inferential statistics then becomes "does this difference represent a reliable and meaningful difference or is it due purely to chance variation and therefore without consistency?" A prime function of inferential statistics is to provide *rigorous* and *logically sound* procedures for answering these questions.

1.2.2   Types of Hypotheses

Before discussing the procedures underlying hypothesis testing, it is necessary to distinguish between *research hypothesis* and *null hypothesis*.

Research hypothesis. The hypothesis derived from the researcher's theory about some social phenomenon is called a research hypothesis. Researchers usually believe that their research hypotheses are true or that they are accurate statements about the condition of the things that they are investigating. They believe that their hypotheses are true to the extent that the theory from which they were derived is adequate. However, theories are only suppositions about the true nature of things, and, thus, hypotheses derived from theories must also be regarded as just tentative suppositions about things until they have been tested. Testing hypotheses means subjecting them to *confirmation* or *disconfirmation*.

Null hypothesis. A null hypothesis is, in a sense, the reverse of research hypothesis. It is also a statement about the reality of things, except that it serves to *deny* what is explicitly indicated in a given research hypothesis. For example, if the researcher states as his or her research hypothesis that "the average grade of Central Queensland University psychology students is 90%," he or she may also state a null hypothesis that can be used to evaluate the accuracy of the research hypothesis. The null hypothesis would be "the average grade of Central Queensland University psychology students is *not* 90%" If the researcher is able to demonstrate that the average grade of these students is at or near 90%, then the researcher concludes that the null hypothesis is refuted or regarded as not true. If the researcher rejects the null hypothesis, then logically the statement that "the average grade of CQU psychology students is 90%" is supported. In other words, the researcher constructs a situation that contains two contradictory statements, namely "the average grade of CQU psychology students is/is not 90%." These statements are worded in such a way that they are mutually exclusive; that is, the confirmation of one is the denial or refutation of the other. *Both cannot coexist simultaneously.* In social research, when we test a hypothesis, we test the null hypothesis — the statement that there is no difference between groups or no relationship between variables. Whether our research hypothesis is supported or refuted depends on the outcome of the test of the null hypothesis.

1.2.3 Testing Hypotheses

Testing hypotheses means subjecting them to some sort of empirical scrutiny to determine if they are supported or refuted by what the researcher observes. Suppose a researcher has collected data pertaining to his or her hypotheses and recorded what was found. The question then is, what are the bases upon which the researcher concludes that his or her hypotheses are supported or refuted? What would prevent two researchers working independently to arrive at quite different conclusions pertaining to the identical information studied? To avoid or at least reduce the amount of subjectivity that exists when social scientists interpret what they have found, they employ a decision rule that specifies the conditions under which the researchers will decide to refute or support the hypotheses they are testing. This decision rule is called the *level of significance.*

1.2.4   Level of Significance

When a difference in characteristics (e.g., IQ, verbal test, personality traits, etc.) between two groups is observed, at what point do we conclude that the difference is a *significant* one? We are usually going to observe differences between various people regarding commonly held characteristics, but how do we decide that the differences mean anything important to us? That is, how do we judge these differences? What may be a significant difference to one researcher may not be considered as such by another. In order that we may introduce greater objectivity into our interpretations of observations, we establish the level of significance. To state a level of significance is to state a probability level at which the researcher will decide to accept or reject the null hypothesis. How do levels of significance operate to enable researchers to make decisions about their observations? To answer this question, we need to look at *probability theory.*

Probability in social research is concerned with the possible outcomes of experiments, that is, the likelihood that one's observations or results are expected or unexpected. For example, if we were to flip a coin 100 times, we would expect that we would get 50 heads and 50 tails. This expected outcome assumes that the coin is not biased. A biased coin, however, would reflect a greater proportion of either heads or tails compared to an unbiased one. Now, if we actually flip the coin 100 times, does the distribution of heads and tails differ from what would be expected? Determining what is expected is based on the possible outcomes associated with our observations. In the flip of a coin, there are only two possible outcomes: head or tail. If the coin is unbiased, then we would expect head to come up 50% of the time and tail to come up 50% of the time. In other words, the probability of getting a head is $P = 0.5$, and the probability of getting a tail is $P = 0.5$. This is what would be expected from flipping an unbiased coin. However, if we flip the coin 100 times and it comes up head 60 times and tail 40 times, will we say that this distribution is so different from our expected distribution as to conclude that the coin is biased? How about an outcome of 70 heads and 30 tails, or 75 heads and 25 tails? The question here is, at what point do we decide to regard an outcome as significantly different from what we would expect according to probability? In social research, we set the probability or significance level at ≤ 0.05. In other words, using our coin example, only when the coin comes up heads (or tails) 95 times or more out of 100 throws, then we say the coin is biased. In statistical terms, we say that the observed outcome is *statistically significant* if the probability of the difference occurring by chance is less than five times out of a hundred (i.e., we conclude that something else other than chance has affected the outcome).

1.2.5   Type I and Type II Errors

Type I error. We use the level of significance to help us decide whether to accept or reject the null hypothesis. In the coin example, when we set the level of significance at 0.05, we will only reject the null hypothesis of neutrality if the coin turns up heads 95 times or more (or tails 5 times or less) in 100 throws. If, for example, the coin turns up heads 96 times out of 100 throws, and we reject the null hypothesis of neutrality, is it not still possible that we could be wrong in rejecting the null hypothesis? Is it not possible that we have, in fact, obtained a statistically rare occurrence by chance? The answer to this question must be "yes."

If we reject the null hypothesis when it is true and should not be rejected, we have committed a *Type I error.* In testing a hypothesis, the level of significance set to decide whether to accept or reject the null hypothesis is the amount of Type I error the researcher is willing to permit. When we employ the 0.05 level of significance, approximately 5% of the time we will be wrong when we reject the null hypothesis and assert its alternative. Then, it would seem that to reduce this type of error, we should set the rejection level as low as possible. For example, if we were to set the level of significance at 0.001, we would risk a Type I error only about one time in every thousand. However, the lower we set the level of significance, the greater is the likelihood that we will make a Type II error.

Type II error. If we fail to reject the null hypothesis when it is actually false, we have committed a Type II error. This type of error is far more common than a Type I error. For example, in the coin experiment, if we set the level of significance at 0.01 for rejecting the null hypothesis and the coin turns up heads 98 times out of 100 throws, the researcher will not be able to reject the null hypothesis of neutrality. That is, based on these observations, the researcher cannot claim that the coin is biased (the alternative hypothesis) even though it may very well be. Only if the coin turns up heads 99 times or more will the researcher be able to reject the null hypothesis.

Then it is clear that Type I and Type II errors cannot be eliminated. They can be minimized, but minimizing one type of error will increase the probability of committing the other error. The lower we set the level of significance, the lesser is the likelihood of a Type I error and the greater the likelihood of a Type II error. Conversely, the higher we set the level of significance, the greater the likelihood of a Type I error and the lesser the likelihood of a Type II error.

1.3 Test Selection

When a researcher is ready to test a specific hypothesis generated from a theory or to answer a research question posed, he or she is faced with the task of choosing an appropriate statistical procedure. Choosing an appropriate statistical procedure to test a specific hypothesis is important for a number of reasons. First, for any hypothesis posited, the statistical procedure chosen must offer a legitimate test of the hypothesis; otherwise, no meaningful interpretation of the results can be made. For example, if a study hypothesizes *differences* in mean scores between groups, it will make no sense for the researcher to choose a test of *relationship* between pairs of variables. Second, choosing an inappropriate statistical procedure can also mean choosing a less-than-robust test that fails to detect significant differences between group scores or significant relationships between pairs of variables (i.e., increasing the probability of committing a Type II error).

Although the researcher is faced with a multitude of statistical procedures to choose from, the choice of an appropriate statistical test is generally based on just two primary considerations: (1) the nature of the hypothesis and (2) the levels of measurement of the variables to be tested.

1.3.1   The Nature of Hypotheses: Test of Difference vs. Test of Relationship

In choosing an appropriate statistical test, the first issue that the researcher must consider is the nature of the hypothesis. Is the intention of the hypothesis to test for differences in mean scores between groups, or is it testing for relationships between pairs of variables?

1.3.1.1   Test of Differences

Testing for differences means that the researcher is interested in determining whether differences in mean scores between groups are due to chance factors or to real differences between the groups as a result of the study's experimental treatment. Suppose, for example, a company whose main business is to help people increase their level of IQ wants to market a new drug called IQADD that is believed to raise IQ scores. To test the effectiveness of this drug, the company conducts an experiment in which 64 random persons are randomly assigned to two groups of n = 34 persons each. The first group is then administered a specific dosage of IQADD, whereas the second group is given a placebo. The company then measures the IQ scores (obtained from a standardized IQ test) for all 64 people to see which group has the higher mean score. Suppose that the IQADD group has a mean IQ score of 134 and the placebo group has a mean IQ score of 110. The question that this company must now ask is whether this difference of 24 IQ points in favor of the IQADD group is big enough to rule out the element of chance as an explanation (and, therefore, must be due to the effect of the IQADD drug), or whether it is so small that chance variation in IQ between the two groups is still a legitimate explanation. To answer this question, the company must choose a statistical test that will appropriately test for the *difference* in mean IQ scores between the two groups.

1.3.1.2 Test of Relationships

Testing of relationships among two or more variables involves asking the question, "Are variations in variable X associated with variations in variable Y?" For example, do students who do well in high school also perform well in university? Do parents with high intelligence tend to have children of high intelligence? Is there a relationship between the declared dividend on stocks and their paper value in the exchange? Is there a relationship between socioeconomic class and recidivism in crime? All these questions concern the relationships among variables, and to answer these questions, researchers must choose statistical tests that will appropriately test for the *relationships* among these variables.

1.3.2 Levels of Measurement

In addition to considering the nature of the hypothesis to be tested (differences or relationships), the researcher must also consider the measurements of the variables to be tested. This is because the levels at which the variables are measured determine the statistical test that is used to analyze the data. Most typically, variables in the behavioral sciences are measured on one of four scales: *nominal, ordinal, interval,* or *ratio* measurements. These four types of scales differ in the number of the following attributes they posses: *magnitude, an equal interval between adjacent units,* and *an absolute zero point.*

Nominal scale. This is the lowest level of measurement and involves simply categorizing the variable to be measured into one of a number of discrete categories. For instance, in measuring "ethnic origin," people may be categorized as American, Chinese, Australian, African, or Indian. Once people have been categorized into these categories, all people in the same category (e.g., those categorized as Americans) are equated on the measurement obtained, even though they may not possess the same amount of the characteristics. Numbers can be assigned to describe the categories, but the numbers are only used to name/label the categories. They have no quantitative value in terms of magnitude.

Ordinal scale. This level of measurement involves ordering or ranking the variable to be measured. For example, people may be asked to rank order four basketball teams according to their skills. Thus, the rank of 1 is assigned to the team that is the most skillful, the rank of 2 to the team that exhibits the next greatest amount of skill, and so forth. These numbers allow the researcher to quantify the magnitude of the measured variable by adding the arithmetic relationships "greater than" and "less than" to the measurement process. Although ordinal scales allow one to differentiate between rankings among the variables being measured, they do not permit the determination of how much of a real difference exists in the measured variable between ranks. The basketball team that is ranked 1 is considered to be better than the team that is ranked 2, but the rankings provide no information as to *how much better* the first-ranked basketball team is. In other words, the intervals between the ranks are not meaningful.

Interval scale. This level of measurement involves being able to specify how far apart two stimuli are on a given dimension. On an ordinal scale, the difference between the basketball team ranked first and the team ranked second does not necessarily equal the distance between teams ranked third and fourth. On an interval scale, however, differences of the same numerical size in scale values *are* equal. For example, on a standardized intelligence measure, a difference of 10 points in IQ scores has the same meaning anywhere along the scale. Thus, the difference in IQ test scores between 80 and 90 is the same as the difference between 110 and 120. However, it would not be correct to say that a person with an IQ score of 100 is *twice* as intelligent as a person with a score of 50. The reason for this is that scales of intelligence tests (and other similar interval scales) do not have a true *zero* that represents a complete absence of intelligence.

Ratio scale. This level of measurement replaces the arbitrary zero point of the interval scale with a true zero starting point that corresponds to the absence of the variable being measured. Thus, with a ratio scale, it is possible to state that a variable has twice, half, or three times as much of the variable measured than another. Take weight as an example. Weight has a true zero point (a weight of zero means that the object is weightless), and the intervals between the units of measurement are equal. Thus, the difference between 10 g and 15 g is equal to the difference between 45 g and 50 g, and 80 g is *twice* as heavy as 40 g.

1.3.3   Choice of Test

Once the researcher has decided on the nature of the hypothesis to be tested (*test of difference* or *test of relationship*) and the levels of measurement of the variables to be included in the analysis (*nominal, ordinal, interval,* or *ratio*), the next step is to choose an appropriate statistical test for analyzing the data. Table 1.1 presents a "Test Selection Grid" that will aid in the selection of an appropriate statistical procedure. From the grid, the researcher chooses a test by considering (1) the purpose of the statistical analysis, (2) the levels of measurement of the variables, (3) the number of sets of scores to be included in the analysis, and (4) whether the sets of scores are related or independent. It should be noted that the grid is restricted to only a number of the most commonly used procedures. There are far more statistical procedures available in SPSS for Windows.

TABLE 1.1

Test Selection Grid

	Relationship	One Set of Scores	Related Two Sets	Independent Two Sets	More than Two Sets
Nominal	Point biserial (rpb) (true/continuous)	Single variable Chi-square (χ^2) test	McNemar significance of change χ^2	Chi-square test of association	Chi-square goodness of fit
	Biserial (rb) (artificial/ continuous)				
	Phi (rΦ) (true/true)				
	Tetrachoric (rt) (artificial/artificial)				
Ordinal	Spearman's rho	Kolmogorov –Smirnov test for ranked data	Wilcoxon matched- pairs signed- ranks test	Mann– Whitney U test	Kruskal– Wallis test
Interval/ Ratio	Pearson's product- moment correlation	One-sample t-test	Related samples t-test	Independent samples t-test	One-way ANOVA (independent)
	Linear regression				Factorial ANOVA (independent)
	Multiple regression				Multivariate ANOVA (related)

2

Introduction to SPSS

2.1 Introduction

SPSS, which stands for *Statistical Product and Service Solutions* (formerly *Statistical Package for the Social Sciences*), is an integrated system of computer programs designed for the analysis of social sciences data. It is one of the most popular of the many statistical packages currently available for statistical analysis. Its popularity stems from the following features of the program:

- It allows for a great deal of flexibility in the data format.
- It provides the user with a comprehensive set of procedures for data transformation and file manipulation.
- It offers the researcher a large number of statistical analyses processes commonly used in social sciences.

For both beginners and advanced researchers, SPSS is an indispensable tool. Not only is it an extremely powerful program, it is also relatively easy to use once the researcher has been taught the rudiments. The Windows version of SPSS has introduced a *point-and-click* interface that allows the researcher to merely point-and-click through a series of windows and dialog boxes to specify the kind of analysis required and the variables involved. This method eliminates the need to learn the very powerful syntax or command language used to execute SPSS analyses (in the older versions that ran on MS-DOS) and has proven to be highly popular for those researchers with little or no interest in learning the *syntax method*. Nevertheless, SPSS for Windows has retained the syntax method, which permits the researcher to execute SPSS analyses by typing commands.

A question that is often asked by the beginner researcher is, "Which method of running SPSS is better?" Both the Windows method and the syntax method have their advantages and disadvantages, and these will be discussed in Section 2.3.

This chapter has been written with the beginner student and researcher in mind, and provides an overview of the two most basic functions of SPSS: (1) how to set up data files in SPSS for Windows, and (2) how to conduct SPSS analysis via the Windows method and the syntax method.

2.2 Setting up a Data File

Suppose a survey has been conducted to investigate the extent to which Australians agree with increases in government spending in the three areas of defense, social security, and child care services. Responses to these three issues were obtained from the questionnaire presented in Table 2.1.

2.2.1 Preparing a Code Book

Prior to data entry, it will be useful to prepare a code book that contains the names of the variables in the questionnaire, their corresponding SPSS variable names, and their coding instructions. An important purpose of such a code book is to allow the researcher to keep track of all the variables in the survey questionnaire and the way they are defined in the SPSS data file. Table 2.2 presents the code book for the questionnaire presented earlier.

TABLE 2.1

Survey Questionnaire

a) Gender 1. _____ Male 2. _____ Female
b) Age _____ (in years)
c) The following three statements relate to increases in government spending in the areas of defense, social security, and child care services. Please consider these three statements carefully and then decide your level of agreement with the government's decision to increase spending. Please indicate your level of agreement by circling the number on each six-point scale.

(i) Increased spending in defence.

1 _____ 2 _____ 3 _____ 4 _____ 5 _____ 6 _____

| Strongly disagree | Moderately disagree | Barely disagree | Barely agree | Moderately agree | Strongly agree |

(ii) Increased spending in social security.

1 _____ 2 _____ 3 _____ 4 _____ 5 _____ 6 _____

| Strongly disagree | Moderately disagree | Barely disagree | Barely agree | Moderately agree | Strongly agree |

(iii) Increased spending in child care services.

1 _____ 2 _____ 3 _____ 4 _____ 5 _____ 6 _____

| Strongly disagree | Moderately disagree | Barely disagree | Barely agree | Moderately agree | Strongly agree |

TABLE 2.2

Code Book

Variable	SPSS Variable Name	Code
Gender	Gender	1 = male
		2 = female
Age	Age	Age in years
Defence	Defence	1 = strongly disagree
		2 = moderately disagree
		3 = barely disagree
		4 = barely agree
		5 = moderately agree
		6 = strongly agree
Social security	Social	1 = strongly disagree
		2 = moderately disagree
		3 = barely disagree
		4 = barely agree
		5 = moderately agree
		6 = strongly agree
Child care services		1 = strongly disagree
		2 = moderately disagree
		3 = barely disagree
		4 = barely agree
		5 = moderately agree
		6 = strongly agree

TABLE 2.3

Raw Data

Gender	Age	Defence	Social	Child
1	24	4	2	1
1	18	5	1	4
2	33	2	5	6
1	29	5	3	4
2	26	3	5	5
2	19	2	5	2
1	36	4	4	3
2	34	3	6	6
1	20	3	5	1
2	21	2	5	3

2.2.2 Data Set

Table 2.3 presents the responses obtained from a sample of ten respondents to the questionnaire.

2.2.3 Creating an SPSS Data File

The following steps demonstrate how the data presented in Table 2.3 are entered into an SPSS data file.

1. When the SPSS program is launched, the following window will open:

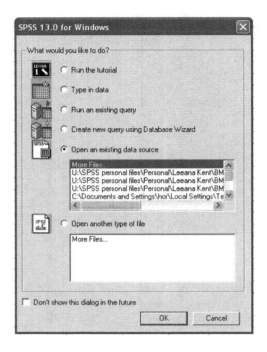

Because the purpose of the present exercise is to create a new data file, close this window by clicking 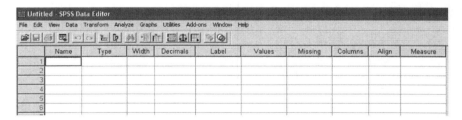 . The following **Untitled – SPSS Data Editor** screen will then be displayed:

2. Prior to data entry, the variables in the data set must be named and defined. In the **Untitled – SPSS Data Editor** screen, the names of the variables are listed down the side, with their characteristics listed along the top (Type, Width, Decimals, Label, Values, Missing, Columns, Align, and Measure). The code book presented in Table 2.2 will serve as a guide in naming and defining the variables. For example, the first variable is named GENDER and is coded 1 = male

and 2 = female. Thus, in the first cell under **Name** in the **Data Editor** screen, type in the name GENDER. To assign the coded values (1 = male, 2 = female) to this variable, click the corresponding cell under **Values** in the **Data Editor** screen. The cell will display

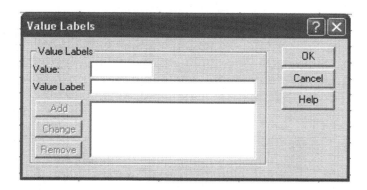

. Click the shaded area to open the following **Value Labels** window:

3. To define the code for male respondents, type **1** in the **Value** cell, and type **Male** in the **Value Label** cell. Next, click [Add] to complete the coding for the male respondents. For female respondents, type **2** in the **Value** cell, and type **Female** in the **Value Label** cell. Next, click [Add] to complete the coding for the female respondents. The completed **Value Labels** window is presented in the following.

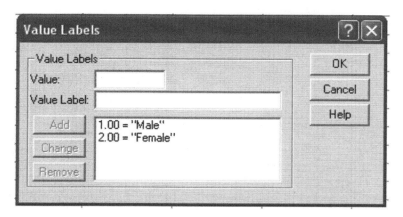

Next, click [OK] to complete the coding for the GENDER variable and to return to the **Untitled – SPSS Data Editor** screen shown in the following.

4. Repeat this coding procedure for the rest of the variables in the code book. Please note that the AGE variable is a *continuous* variable and, therefore, has no coded values.

5. If the researcher wishes to attach a label to a variable name (to provide a longer description for that variable), this can be done by typing a label in the corresponding cell in the **Label** column. For example, the researcher may wish to attach the label **increased spending on defense** to the variable DEFENCE. This label will be printed in the analysis output generated by SPSS. The following **Untitled – SPSS Data Editor** screen displays the names of all the variables listed in the code book, and where relevant, their **Labels** and **Values** codes.

	Name	Type	Width	Decimals	Label	Values	Missing
1	GENDER	Numeric	8	2		{1.00, MALE}.	None
2	AGE	Numeric	8	2		None	None
3	DEFENCE	Numeric	8	2	increased spending on defence	{1.00, strongly	None
4	SOCIAL	Numeric	8	2	increased spending on social security	{1.00, strongly	None
5	CHILD	Numeric	8	2	increased spending on child-care services	{1.00, strongly	None

2.2.4   Data Entry

Data can only be entered via the **Data View** screen. Switch the present **Variable View** to **Data View** by clicking the **Data View** tab

[\ **Data View** ⟨ Variable View /]   at the bottom left-hand corner of the screen. In the **Data View** screen, the rows represent the respondents, and the columns represent the variables. Beginning with the first data cell (row 1, column 1), type in the data presented in Table 2.3. The following **Data View** screen shows the data obtained from the ten respondents.

2.2.5   Saving and Editing a Data File

Once data entry is completed, the data file can be saved. From the menu bar, click **File**, then **Save As**. Once it has been decided where the data file is to be saved to, type a name for the file. As this is a data file, SPSS will automatically append the suffix **.SAV** to the data file name (e.g., **TRIAL.SAV**). To edit an existing file, click **File**, then **Open**, and then **Data** from the menu

	GENDER	AGE	DEFENCE	SOCIAL	CHILD
1	1.00	24.00	4.00	2.00	1.00
2	1.00	18.00	5.00	1.00	4.00
3	2.00	33.00	2.00	5.00	6.00
4	1.00	29.00	5.00	3.00	4.00
5	2.00	26.00	3.00	5.00	5.00
6	2.00	19.00	2.00	5.00	2.00
7	1.00	36.00	4.00	4.00	3.00
8	2.00	34.00	3.00	6.00	6.00
9	1.00	20.00	3.00	5.00	1.00
10	2.00	21.00	2.00	5.00	3.00

bar. Scroll through the names of the data files and double-click on the data file to open it.

2.3   SPSS Analysis: Windows Method vs. Syntax Method

Once the SPSS data file has been created, the researcher can conduct the chosen analysis either through the Windows method (point-and-click) or the syntax method. The primary advantage of using the Windows method is clearly its ease of use. With this method, the researcher accesses the pull-down menu by clicking **Analyze** in either the **Data View** or **Variable View** mode, and then point-and-clicks through a series of windows and dialog boxes to specify the kind of analysis required and the variables involved. There is no need to type in any syntax or commands to execute the analysis. Paradoxically, although this procedure seems ideal at first, it is not always the method of choice for the more advanced and sophisticated users of the program. Rather, there is clearly a preference for the syntax method among these users. This preference stems from several good reasons from learning to use the syntax method.

First, when conducting complex analysis, the ability to write and edit command syntax is advantageous. For example, if a researcher misspecifies a syntax command for a complex analysis and wants to go back and rerun it with minor changes, or if the researcher wishes to repeat an analysis multiple times with minor variations, it is often more efficient to write and edit the syntax command directly than to repeat the Windows pull-down menu sequences. Second, from my teaching experience with SPSS, I believe that students have a better "feel" for statistics if they have to write syntax commands to generate the specific statistics they need, rather than merely relying on pull-down menus. In other words, it provides a better learning experience. Finally, and perhaps most important, several SPSS procedures are available only via the syntax method.

2.3.1   SPSS Analysis: Windows Method

Once the data have been entered, the researcher can begin the data analysis. Suppose the researcher is interested in obtaining general **descriptive statistics** for all of the variables entered in the data set TRIAL.SAV (see Subsection 2.2.5).

1. From the menu bar, click **Analyze**, then **Descriptive Statistics**, and then **Frequencies.** The following **Frequencies** window will open:

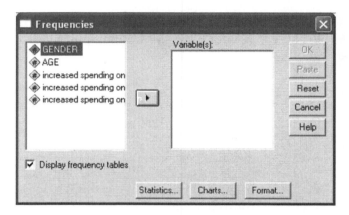

2. In the left-hand field containing the study's five variables, click (highlight) these variables, and then click 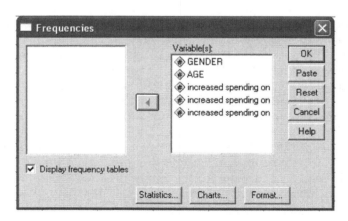 to transfer the selected variables to the **Variable(s)** field.

3. Click [Statistics...] to open the **Frequencies: Statistics** window. Suppose the researcher is only interested in obtaining statistics for the **Mean, Median, Mode,** and **Standard Deviation** for the five variables, check the fields related to these statistics in the **Frequencies: Statistics** window. Next click [Continue] .

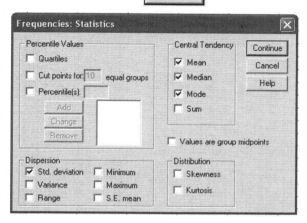

4. When the **Frequencies** window opens, run the analysis by clicking [OK]. See Table 2.4 for the results.

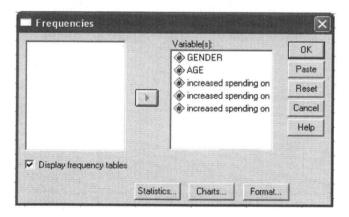

2.3.2   SPSS Analysis: Syntax Method

1. From the menu bar, click **File**, then **New**, and then **Syntax**. The following **Syntax1 – SPSS Syntax Editor** window will open.

2. Type the **Frequencies** analysis syntax command in the **Syntax1 – SPSS Syntax Editor** window. If the researcher is interested in

obtaining all descriptive statistics (and not just the mean, median, mode, and standard deviation), then replace the syntax: /STATISTCS=MEAN MEDIAN MODE STDDEV. with /STATIS-TICS=ALL.

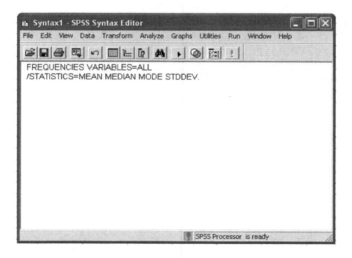

3. To run the Frequencies analysis, click **Run** and then **All**.

(**Note: The Appendix presents a summary of the SPSS syntax files employed for all the examples in this book.**)

2.3.3 SPSS Output

Frequencies

TABLE 2.4

Frequencies Output

Statistics

	GENDER	AGE	Increased Spending on Defence	Increased Spending on Social Security	Increased Spending on Child Care Services
N Valid	10	10	10	10	10
Missing	0	0	0	0	0
Mean	1.5000	26.0000	3.3000	4.1000	3.5000
Median	1.5000	25.0000	3.0000	5.0000	3.5000
Mode	1.00[a]	18.00[a]	2.00[a]	5.00	1.00[a]
Standard deviation	0.52705	6.66667	1.15950	1.59513	1.84089

[a] Multiple modes exist. The smallest value is shown.

Frequency Table

Gender

		Frequency	Percentage	Valid Percentage	Cumulative Percentage
Valid	MALE	5	50.0	50.0	50.0
	FEMALE	5	50.0	50.0	100.0
	Total	10	100.0	100.0	

Age

		Frequency	Percentage	Valid Percentage	Cumulative Percentage
Valid	18.00	1	10.0	10.0	10.0
	19.00	1	10.0	10.0	20.0
	20.00	1	10.0	10.0	30.0
	21.00	1	10.0	10.0	40.0
	24.00	1	10.0	10.0	50.0
	26.00	1	10.0	10.0	60.0
	29.00	1	10.0	10.0	70.0
	33.00	1	10.0	10.0	80.0
	34.00	1	10.0	10.0	90.0
	36.00	1	10.0	10.0	100.0
	Total	10	100.0	100.0	

Increased Spending on Defence

		Frequency	Percentage	Valid Percentage	Cumulative Percentage
Valid	Moderately disagree	3	30.0	30.0	30.0
	Barely disagree	3	30.0	30.0	60.0
	Barely agree	2	20.0	20.0	80.0
	Moderately agree	2	20.0	20.0	100.0
	Total	10	100.0	100.0	

Increased Spending on Social Security

		Frequency	Percentage	Valid Percentage	Cumulative Percentage
Valid	Strongly disagree	1	10.0	10.0	10.0
	Moderately disagree	1	10.0	10.0	20.0
	Barely disagree	1	10.0	10.0	30.0
	Barely agree	1	10.0	10.0	40.0
	Moderately agree	5	50.0	50.0	90.0
	Strongly agree	1	10.0	10.0	100.0
	Total	10	100.0	100.0	

Increased Spending on Child Care Services

		Frequency	Percentage	Valid Percentage	Cumulative Percentage
Valid	Strongly disagree	2	20.0	20.0	20.0
	Moderately disagree	1	10.0	10.0	30.0
	Barely disagree	2	20.0	20.0	50.0
	Barely agree	2	20.0	20.0	70.0
	Moderately agree	1	10.0	10.0	80.0
	Strongly agree	2	20.0	20.0	100.0
	Total	10	100.0	100.0	

2.3.4 Results and Interpretation

The **Statistics** table presents the requested mean, median, mode, and standard deviation statistics for the five variables. The **Gender** variable is a nominal (categorical) variable, and, as such, its mean, median, and standard deviation statistics are not meaningful. The remaining four variables of **Age**, **Defence**, **Social**, and **Child** are measured at least at the ordinal level (i.e., they are continuous variables), and, as such, their mean, median, and standard deviation statistics can be interpreted.

The results presented in the **Statistics** table show that the ten respondents in the survey have a mean age of 26 yr and a median age of 25 yr. Because there is no one age that occurs more frequently than others, SPSS presents the lowest age value of 18 as the mode. For the three variables of "support for increased spending" on defence, social security, and child care services, the results show that support for increased spending for social security is highest (Mean = 4.10; Median = 5.00), followed by child care services (Mean = 3.50; Median = 3.50), and defence (Mean = 3.30; Median = 3.00). The results also show that the variables of **defence** and **child** have multiple modes, and, as such, SPSS has presented their lowest values (defence: Mode = 2.00; child: Mode = 1.00). The variable of **social** has a single mode of 5.00.

For the age variable, the standard deviation shows that its average deviation (dispersion) from the mean is 6.66 yr. For the three variables of defence, social, and child, the results show that support for increased spending on child care services has the largest average variation (SD = 1.84) from its mean score. The standard deviation scores of support for increased spending for defence (SD = 1.59) and social security (SD = 1.59) are similar.

The **Frequency Table** presents the breakdown of the frequency distributions for the five variables (Gender, Age, Defence, Social, and Child). For each variable, the Frequency Table presents (1) the **frequency** of occurrence for each value within that variable, (2) the frequency for each value expressed as a **percentage** of the total sample, (3) the **valid percentage** for each value, controlling for missing cases, and (4) the **cumulative percentage** for each succeeding value within that variable. For example, the frequency table for the variable of Gender shows that there are five males and five females in the sample and that these two groupings represent 50% each of the total sample. Because there are no missing cases, the valid percentage values are identical to the percentage values. *If there are missing cases, then the valid percentage values should be interpreted.* The cumulative percentage presents the percentage of scores falling at or below each score. Thus, for the sample of ten respondents, the five males in the sample represent 50% of the sample, and the additional five females represent a cumulative percentage of 100%.

The frequency tables for the variables of age, defence, social, and child care services are interpreted in exactly the same way.

3

Multiple Response

3.1   Aim

MULT RESPONSE analysis allows the researcher to analyze research questions that can have multiple responses. For example, a researcher may ask respondents to *name* all the newspapers read within the last week, or to *circle* all newspapers read within the last week from a list of newspapers. One way to generate descriptive statistics for each of the newspapers nominated is to do a simple FREQUENCIES analysis. However, an ordinary FREQUENCIES analysis will only generate descriptive statistics for each nominated newspaper *separately* (e.g., the number and percentage of respondents who nominated newspaper A). This procedure will not generate statistics on the basis of the entire "group" of newspapers nominated. For example, if the researcher is interested in the number of respondents who nominated newspaper A as a *percentage of the total number of newspapers read*, then the MULT RESPONSE procedure should be used.

3.2   Methods of MULT RESPONSE Procedures

There are two ways to perform a frequency run with multiple response data. Whichever way the researcher chooses, the procedure will involve combining variables into groups. One way to organize multiple-response data is to create, for each possible response, a variable that can have one of two values, such as 1 for *yes* and 2 for *no*; this is the **multiple-dichotomy** method. Alternatively, on the basis of the responses collected from all respondents, the researcher can create variables to represent, for example, all the newspapers read. Each variable (newspaper) will have a value representing that newspaper, such a 1 for *The Australian*, 2 for *Courier Mail*, and 3 for *The Canberra Times*. This is the **multiple-response** method.

3.3   Example of the Multiple-Dichotomy Method

Suppose that in a survey of political party preference, the following question was asked:

"Why do you prefer that political party?" (you can choose more than one reason).

		1. Yes	2. No
1.	The party is honest.	____	____
2.	The party has integrity.	____	____
3.	The party is trustworthy.	____	____
4.	The party has always kept its promises.	____	____
5.	The party has strong leadership.	____	____

In this example, the respondent is asked to endorse all the reasons for preferring a specific political party. Each reason is, therefore, a separate variable with two possible values: 1 for *yes*, and 2 for *no*. The five reasons are named **HONEST, INTEG, TRUST, PROMISE**, and **LEADER** in the data set given in the following. To do a **Multiple-Dichotomy** frequency analysis on the given reasons, choose either the **Windows** method (Subsection 3.3.2) or the **Syntax File** method (Subsection 3.3.3).

3.3.1   Data Entry Format

The data set has been saved under the name **EX3.SAV.**

Variables	Column(s)	Code
SEX	1	1 = male, 2 = female
Party	2	1 = Labor, 2 = Liberal
		3 = National, 4 = Democrat
REASON1	3	1 = The party is honest
		2 = The party has integrity
		3 = The party is trustworthy
		4 = The party has always kept its promises
		5 = The party has strong leadership
REASON2	4	As earlier
REASON3	5	As earlier
HONEST	6	1 = yes, 2 = no
INTEG	7	1 = yes, 2 = no
TRUST	8	1 = yes, 2 = no
PROMISE	9	1 = yes, 2 = no
LEADER	10	1 = yes, 2 = no

3.3.2 Windows Method

1. From the menu bar, click **Analyze**, then **Multiple Response**, and then **Define Sets.** The following **Define Multiple Response Sets** window will open:

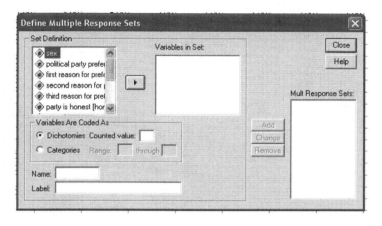

2. In the **Set Definition** field containing the study's variables, click (highlight) the variables (**HONEST, INTEG, TRUST, PROMISE,** and **LEADER)** that will be grouped in the multiple-response set.

   Click 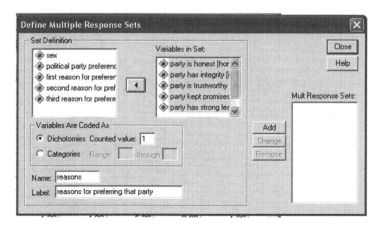 to transfer the selected variables to the **Variables in Set** field. Because only those variables (reasons) that have been coded 1 (for *yes*) will be grouped for analysis, check the **Dichotomous Counted value** field and then type **1** in the field next to it. Next, in the **Name** field, type in a name for this multiple-response set (e.g., *reasons*), and in the **Label** field, type in a label (e.g., *reasons for preferring that party*).

3. Click 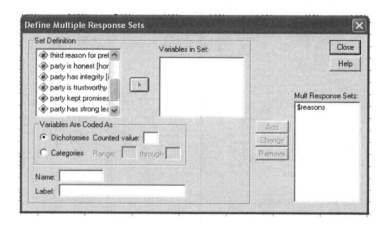 **Add** to transfer this response set to the **Mult Response Sets** field. The grouped response set is given the name *$reasons*.

   Click **Close** to close this window.

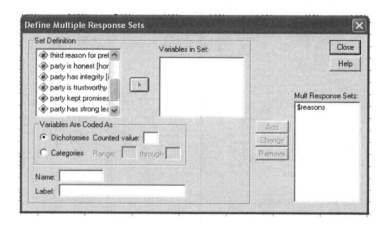

4. From the menu bar, click **Analyze**, then **Multiple Response**, and then **Frequencies.** The following **Multiple Response Frequencies** window will open:

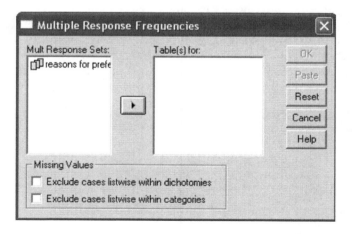

5. Transfer the grouped response set (*reasons for preferring that party*) in the **Mult Response Sets** field to the **Table(s) for** field by clicking (highlight) the response set, and then clicking ▶ .

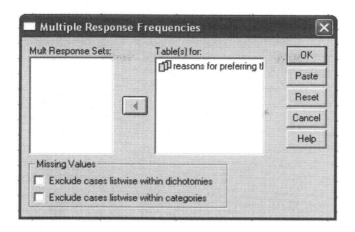

6. Click [OK] to run a multiple-response frequencies analysis for the variables (**HONEST, INTEG, TRUST, PROMISE,** and **LEADER**) in the grouped response set. See Table 3.1 for the results.

3.3.3 SPSS Syntax Method

Multiple-Dichotomy Frequency Analysis
MULT RESPONSE GROUPS=REASONS 'REASONS FOR PREFERRING THAT PARTY' (HONEST TO LEADER(1))/FREQUENCIES=REASONS.

3.3.4 SPSS Output

TABLE 3.1

Multiple Response (Multiple-Dichotomy) Output

Group REASONS FOR PREFERRING THAT PARTY
        (Value tabulated = 1)

Dichotomy label	Name	Count	Pct of Responses	Pct of Cases
party is honest	HONEST	13	23.2	65.0
party has integrity	INTEG	8	14.3	40.0
party is trustworthy	TRUST	10	17.9	50.0
party kept promises	PROMISE	12	21.4	60.0
party has strong leader	LEADER	13	23.2	65.0
		------	-----	-----
Total responses		56	100.0	280.0

0 missing cases; 20 valid cases

3.3.5   Results and Interpretation

In Table 3.1, the **Count** column presents the number of respondents who answered *yes* to each of the five reasons. Thus, of the 20 respondents included in the analysis, 13 endorsed *"party is honest"* as a reason for preferring that political party, 8 endorsed *"party has integrity"* as a reason, 10 endorsed *"party is trustworthy"* as a reason, 12 endorsed *"party kept promises"* as a reason, and 13 endorsed *"party has strong leader"* as a reason. Thus, a total of 56 *yes* responses were generated from the sample of 20 respondents.

The **Pct of Responses** column presents the number of respondents who answered *yes* to each of the five reasons (in the **Count** column) as a percentage of the total number of *yes* responses generated. For example, the 13 respondents who endorsed *"party is honest"* as a reason for preferring that political party represent 23.2% of the total number of *yes* responses (56) generated.

The **Pct of Cases** column presents the number of respondents who answered *yes* to each of the five reasons (in the **Count** column) as a percentage of the total valid sample. For example, the 13 respondents who endorsed *"party is honest"* as a reason represent 65% of the total valid sample (N = 20 cases).

3.4   Example of the Multiple-Response Method

Using the same example as given earlier, the following question was asked. *"Why do you prefer that political party?"*

With the multiple-response method, the grouping of the multiple responses is different from the multiple-dichotomy method. With the multiple-response method, a predetermined list of reasons will be used by the researcher to match the reasons nominated by the respondents. For example, a researcher may have the following list of reasons, for which a numerical value has been assigned to each reason:

1 = The party is honest
2 = The party has integrity
3 = The party is trustworthy
4 = The party has always kept its promises
5 = The party has strong leadership

Suppose that for this particular survey, each respondent is allowed to nominate a maximum of three reasons in response to the earlier question. These responses are labeled **REASON1**, **REASON2**, and **REASON3** in the data set: **EX3.SAV**.

3.4.1 Windows Method

1. From the menu bar, click **Analyze**, then **Multiple Response**, and then **Define Sets**. The following **Define Multiple Response Sets** window will open.

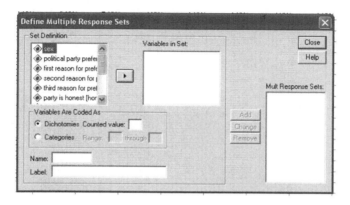

2. In the **Set Definition** field containing the study's variables, click (highlight) the variables (**REASON1, REASON2,** and **REASON3**) that will be grouped in the multiple-response set. Click 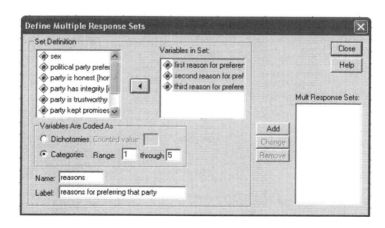 to transfer the selected variables to the **Variables in Set** field. Because these three variables (reasons) have been coded from a predetermined list of five reasons (see Section 3.4), check the **Categories** field, and type **1** through **5** in the **Range** fields. Next, in the **Name** field, type in a name for this multiple-response set (e.g., *reasons*), and in the **Label** field, type in a label (e.g., *reasons for preferring that party*).

Click [Add] to transfer this response set to the **Mult Response Sets** field. The grouped response set is given the name *$reasons*.

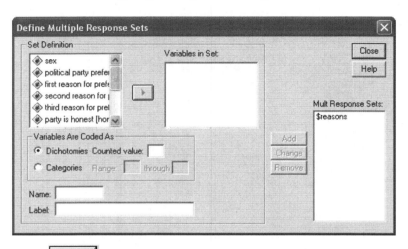

Click [Close] to close this window.

3. From the menu bar, click **Analyze**, then **Multiple Response**, and then **Frequencies.** The following **Multiple Response Frequencies** window will open.

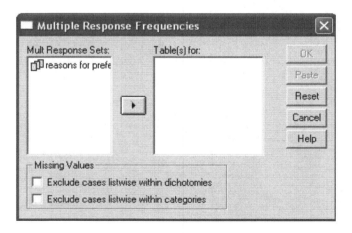

4. Transfer the grouped response set (*reasons for preferring that party*) in the **Mult Response Sets** field to the **Table(s) for:** cell by clicking

   the response set, and then clicking [►] .

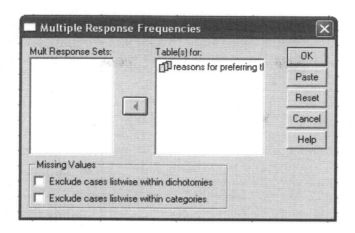

5. Click [OK] to run a multiple-response frequencies analysis for
   the variables (REASON1, REASON2, and REASON3) in the grouped
   response set. See Table 3.2 for the results.

3.4.2   SPSS Syntax Method

MULT RESPONSE GROUPS=REASONS 'REASONS FOR PREFER-
   RING THAT PARTY' (REASON1 TO REASON3 (1,5))
/FREQUENCIES=REASONS.

3.4.3   SPSS Output

TABLE 3.2

Multiple Response (Multiple-Response) Output

```
Group REASONS FOR PREFERRING THAT PARTY
```

Category label	Code	Count	Pct of Responses	Pct of Cases
party is honest	1	11	18.3	55.0
party has integrity	2	12	20.0	60.0
party is trustworthy	3	12	20.0	60.0
party has kept its promises	4	11	18.3	55.0
party has strong leadership	5	14	23.3	70.0
		-------	-----	-----
Total responses		60	100.0	300.0

```
0 missing cases; 20 valid cases
```

3.4.4   Results and Interpretation

In Table 3.2, the **Count** column presents the number of respondents who nominated each of the five reasons. Thus, of the 20 respondents included in the analysis, 11 nominated *"party is honest"* as a reason for preferring that political party, 12 nominated *"party has integrity"* as a reason, 12 nominated *"party is trustworthy"* as a reason, 11 nominated *"party kept promises"* as a reason, and 14 nominated *"party has strong leader"* as a reason. Thus, a total of 60 responses were generated from the sample of 20 respondents.

The **Pct of Responses** column presents the number of respondents who nominated each of the five reasons (in the **Count** column) as a percentage of the total number of responses generated. For example, the 13 respondents who nominated *"party is honest"* as a reason, represent 18.3% of the total number of responses (60) generated.

The **Pct of Cases** column presents the number of respondents who nominated each of the five reasons (in the **Count** column) as a percentage of the total valid sample. For example, the 11 respondents who nominated *"party is honest"* as a reason, represent 55% of the total valid sample (N = 20 cases).

3.5   Cross-Tabulations

Cross-tabulations can be produced by MULT RESPONSE. Both individual and group variables can be tabulated together. Using the earlier example, suppose the researcher wants to cross-tabulate the variable *reasons for preferring that party* with the between-groups variable of SEX (coded 1 = male, 2 = female). The **Multiple-Dichotomy** method will be used to demonstrate this example. The same data set will be used for this example: **EX3.SAV**.

3.5.1   Windows Method

1. From the menu bar, click **Analyze**, then **Multiple Response**, and then **Define Sets.** The following **Define Multiple Response Sets** window will open.

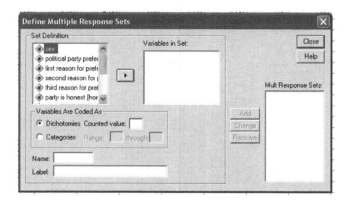

2. In the **Set Definition** field containing the study's variables, click (highlight) the variables (**HONEST, INTEG, TRUST, PROMISE,** and **LEADER)** that will be grouped in the multiple-response set.

Click ▶ to transfer the selected variables to the **Variables in Set** field. Because only those variables (reasons) that have been coded 1 (for *yes*) will be grouped for analysis, check the **Dichotomous Counted value** field and type **1** in the field next to it. Next, type a **Name** for this multiple-response set (e.g., *reasons*), and give it a **Label** (e.g., *reasons for preferring that party*).

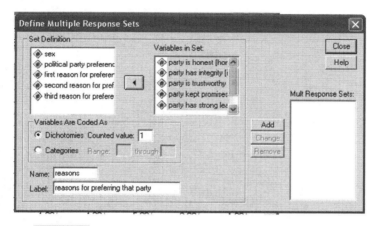

Click **Add** to transfer this response set to the **Mult Response Sets** field. The grouped response set is given the name *$reasons*.

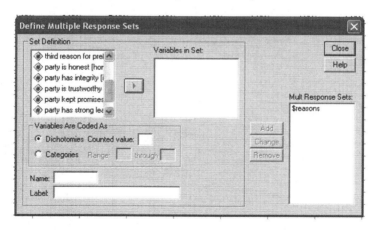

Click **Close** to close this window.

3. From the menu bar, click **Analyze,** then **Multiple Response,** and then **Crosstabs.** The following **Multiple Response Crosstabs** window will open.

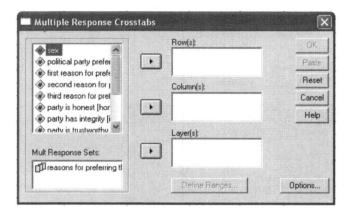

4. Transfer the grouped response set (*reasons for preferring that party*) in the **Mult Response Sets** field to the **Row(s)** field by clicking (highlight) the response set, and then clicking [▶]. Next, transfer the between-groups variables of **SEX** in the **Multiple Response Crosstabs** field to the **Column(s)** field by clicking (highlight) the **SEX** variable, and then clicking [▶].

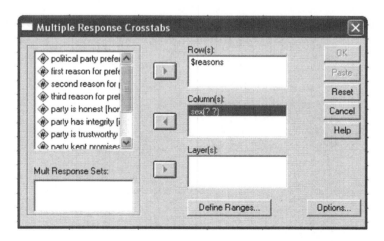

5. Click [Define Ranges...] to define the range for the between-groups variable of **SEX** (coded 1 = male, 2 = female). When the following window opens, type the number **1** in the **Minimum** field and the number **2** in the **Maximum** field, and then click [Continue].

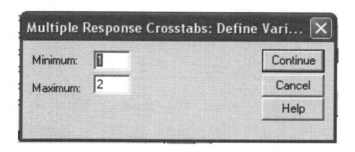

The following **Multiple Response Crosstabs** window will open.

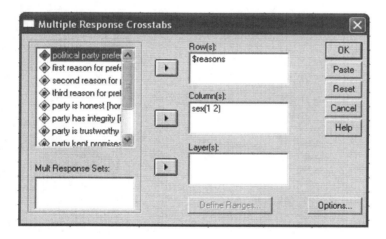

6. Click [Options...] to obtain *row, column,* and *total* percentages in the cross-tabulation table. When the following **Multiple Response Crosstabs: Options** window opens, check the **Row, Column,** and **Total** fields, and then click [Continue].

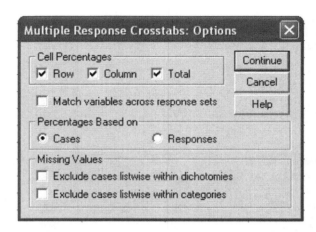

7. When the **Multiple Response Crosstabs** window opens, click
OK to run the cross-tabulation analysis. See Table 3.3 for the results.

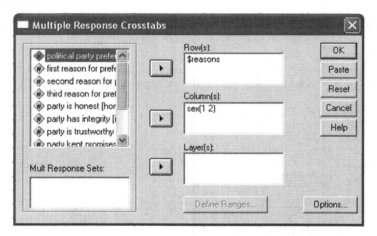

3.5.2 SPSS Syntax Method

MULT RESPONSE GROUPS=REASONS 'REASONS FOR PREFER-RING THAT PARTY' (HONEST TO LEADER(1))
/VARIABLES SEX(1,2)
/TABLES=REASONS BY SEX
/CELLS=ALL.

This syntax file produces the **REASONS*SEX** cross-tabulation table. The **CELLS** syntax produces cell counts, row percentages, column percentages, and two-way table total percentages.

3.5.3 SPSS Output

TABLE 3.3

Multiple Response Cross-Tabulation Output

Reasons by Sex
Multiple Response

Case Summary

		Cases				
	Valid		Missing		Total	
	N	Percentage	N	Percentage	N	Percentage
REASONS*sex	20	100.0%	0	.0%	20	100.0%

REASONS*sex Cross-Tabulation

			Sex		Total
			Male	Female	
Reasons for preferring that party	Party is honest	Count	7	6	13
		% within REASONS	53.8%	46.2%	
		% within sex	63.6%	66.7%	
		% of Total	35.0%	30.0%	65.0%
	Party has integrity	Count	3	5	8
		% within REASONS	37.5%	62.5%	
		% within sex	27.3%	55.6%	
		% of Total	15.0%	25.0%	40.0%
	Party is trustworthy	Count	6	4	10
		% within REASONS	60.0%	40.0%	
		% within sex	54.5%	44.4%	
		% of Total	30.0%	20.0%	50.0%
	Party kept promises	Count	7	5	12
		% within REASONS	58.3%	41.7%	
		% within sex	63.6%	55.6%	
		% of Total	35.0%	25.0%	60.0%
	Party has strong leaders	Count	7	6	13
		% within REASONS	53.8%	46.2%	
		% within sex	63.6%	66.7%	
		% of Total	35.0%	30.0%	65.0%
Total		Count	11	9	20
		% of Total	55.0%	45.0%	100.0%

Note: Percentages and totals are based on respondents.

[a]Dichotomy group tabulated at value 1.

3.5.4 Results and Interpretation

In Table 3.3, **Count** presents the *frequencies* for all the cells. Thus, 7 males and 6 females nominated *"party is honest"* as a reason for preferring that political party, 3 males and 5 females nominated *"party has integrity"* as a reason, 6 males and 4 females nominated *"party is trustworthy"* as a reason, 7 males and 5 females nominated *"party kept promises"* as a reason, and 7 males and 6 females nominated *"party has strong leader"* as a reason.

Row pct (% within REASONS) presents the number of male and female respondents who nominated each of the five reasons (in the **Count** column) as a *percentage* of the number of respondents in each *reason* category. For example, a total of 13 respondents (7 males, 6 females) nominated *"party is honest"* as a reason for preferring that political party. Thus, the 7 males and 6 females represent 53.8% and 46.2% of respondents who nominated that reason, respectively. Similarly, a total of 8 respondents (3 males, 5 females) nominated *"party has integrity"* as a reason for preferring that political party. Thus, the 3 males and 5 females represent 37.5% and 62.5% of respondents who nominated that reason, respectively.

Col pct (% within sex) presents the number of male and female respondents who nominated each of the five reasons (in the **Count** column) as a *percentage* of the total number of male and female respondents, respectively. For example, the total sample consists of 20 respondents (11 males, 9 females). Of the 11 male respondents, 7 nominated *"party is honest"* as a reason for preferring that political party. Similarly, of the 9 female respondents, 6 nominated *"party is honest"* as a reason for preferring that political party. Thus, the 7 males and 6 females represent 63.6% and 66.7% of the total male and female respondents, respectively, who nominated that reason. Similarly, of the 11 male respondents, 3 nominated *"party has integrity"* as a reason for preferring that political party. Of the 9 female respondents, 5 nominated *"party has integrity"* as a reason for preferring that political party. Thus, the 3 males and 5 females represent 27.3% and 55.6% of the total male and female respondents, respectively, who nominated that reason.

Tab pct (% of Total) presents the two-way table total percentages. Thus, for the total sample of 20 respondents, the 7 males who nominated *"party is honest"* represent 35% of the total sample, whereas the 6 females who nominated the same reason represent 30% of the total sample. Similarly, the 3 males who nominated *"party has integrity"* represent 15% of the total sample, whereas the 5 females who nominated the same reason represent 25% of the total sample.

4

T-Test for Independent Groups

4.1 Aim

The independent t-test is used for testing the differences between the means of two independent groups. It is particularly useful when the research question requires the comparison of variables (measured at least at the *ordinal* level) obtained from two independent samples. For example:

- "Do males and females differ in performance on a standardized achievement test?"
- "What is the effect of drug vs. no drug on rats' maze-learning behavior?"
- "Does the recidivism rate of juvenile offenders who are provided with father figures differ from those without father figures?"

4.2 Checklist of Requirements

In any one analysis, there must be:

- Only one independent (grouping) variable (IV) (e.g., subject's gender)
- Only two levels for that IV (e.g., male and female)
- Only one dependent variable (DV)

4.3   Assumptions

- The sampling distribution of differences between means is normally distributed.
- Homogeneity of variance.

4.4   Example

A researcher wants to investigate whether first-year male and female students at a university differ in their memory abilities. Ten students of each group were randomly selected from the first-year enrolment roll to serve as subjects. All 20 subjects were read 30 unrelated words and then asked to recall as many of the words as possible. The numbers of words correctly recalled by each subject were recorded.

	Malés		Females
s1	16	s1	24
s2	14	s2	23
s3	18	s3	26
s4	25	s4	17
s5	17	s5	18
s6	14	s6	20
s7	19	s7	23
s8	21	- s8	26
s9	16	s9	24
s10	17	s10	20

4.4.1   Data Entry Format

The data set has been saved under the name: **EX4.SAV**.

Variables	Column	Code
GENDER	1	1 = male, 2 = female
WORDS	2	Number of words correctly recalled

4.4.2   Windows Method

1. From the menu bar, click **Analyze**, then **Compare Means**, and then **Independent-Samples T-test**. The following window will open:

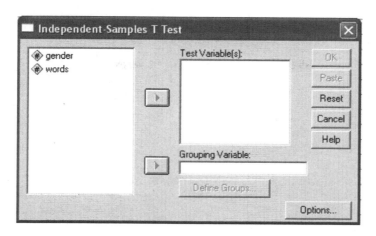

2. Because **GENDER** is the grouping (independent) variable, transfer it to the **Grouping Variable** field by clicking (highlighting) the variable and then clicking [▶]. As **WORDS** is the test (dependent) variable, transfer it to the **Test Variable(s)** field by clicking (highlighting) the variable and then clicking [▶].

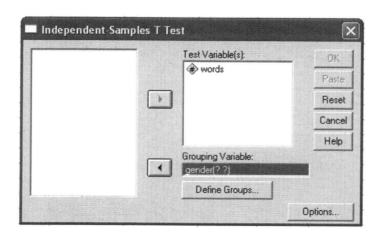

3. Click [Define Groups...] to define the range for the grouping variable of **GENDER** (coded 1 = male, 2 = female). When the following **Define Groups** window opens, type **1** in the **Group 1** field and **2** in the **Group 2** field, and then click [Continue].

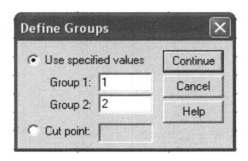

4. When the following **Independent-Samples T Test** window opens, run the T-Test analysis by clicking ![OK]. See Table 4.1 for the results.

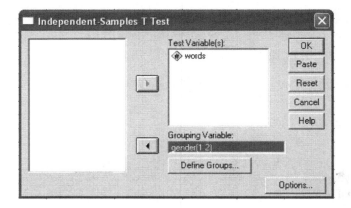

4.4.3   SPSS Syntax Method

T-TEST GROUPS=GENDER(1,2)
/VARIABLES=WORDS.

4.4.4   SPSS Output

TABLE 4.1

Independent T-Test Output

Group Statistics

	GENDER	N	Mean	Standard Deviation	Standard Error Mean
WORDS	MALE	10	17.7000	3.3350	1.0546
	FEMALE	10	22.1000	3.1780	1.0050

Independent Samples Test

		Levene's Test for Equality of Variances		T-Test for Equality of Means					95% Confidence Interval of the Difference	
		F	Sig.	t	df	Sig. (2-tailed)	Mean Difference	Standard Error Difference	Lower	Upper
WORDS	Equal variances assumed	.087	.772	–3.020	18	.007	–4.4000	1.4568	–7.4606	–1.3394
	Equal variances not assumed			–3.020	17.958	.007	–4.4000	1.4568	–7.4611	–1.3389

4.4.5 Results and Interpretation

Levene's Test for Equality of Variances tests the hypothesis that the two population variances are equal. In this example, the Levene statistic is F = 0.087, and the corresponding level of significance is large (i.e., $p > .05$) (see Table 4.1). Thus, the assumption of homogeneity of variance has not been violated, and the **Equal variances assumed** t-test statistic can be used for evaluating the null hypothesis of equality of means. If the significance level of the Levene statistic is small (i.e., $p < .05$), the assumption that the population variances are equal is rejected and the **Equal variances not assumed** t-test statistic should be used.

The result from the analysis indicates that there is a significant difference between the male and female samples in the number of words correctly recalled, t(df = 18) = –3.02, $p < .01$. The mean values indicate that females correctly recalled significantly more words (M = 22.10) than males (M = 17.70).

5

Paired-Samples T-Test

5.1 Aim

The paired-samples t-test is used in **repeated measures** or **correlated groups** design, in which each subject is tested twice on the same variable. A common experiment of this type involves the *before and after* design. The test can also be used for the **matched group** design in which pairs of subjects that are matched on one or more characteristics (e.g., IQ, grades, and so forth) serve in the two conditions. As the subjects in the groups are matched and not independently assigned, this design is also referred to as a **correlated groups** design.

5.2 Checklist of Requirements

- In any one analysis, there must be only two sets of data.
- The two sets of data must be obtained from (1) the same subjects or (2) from two matched groups of subjects.

5.3 Assumptions

- The sampling distribution of the means should be normally distributed.
- The sampling distribution of the difference scores should be normally distributed.

5.4   Example

A researcher designed an experiment to test the effect of drug X on eating behavior. The amount of food eaten by a group of rats in a one-week period prior to ingesting drug X was recorded. The rats were then given drug X, and the amount of food eaten in a one-week period was again recorded. The following amounts of food in grams were eaten during the "before" and "after" conditions.

	Food Eaten	
	Before Ingesting Drug X	After Ingesting Drug X
s1	100	60
s2	180	80
s3	16	110
s4	220	140
s5	140	100
s6	250	200
s7	170	100
s8	220	180
s9	120	140
s10	210	130

5.4.1   Data Entry Format

The data set has been saved under the name **EX5.SAV**.

Variables	Column	Code
BEFORE	1	Food eaten (in grams)
AFTER	2	Food eaten (in grams)

5.4.2   Windows Method

1. From the menu bar, click **Analyze**, then **Compare Means**, and then **Paired-Samples T test**. The following window will open:

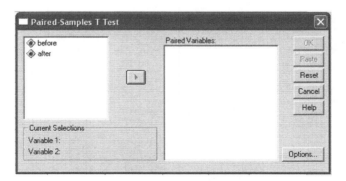

2. Transfer both the **BEFORE** and **AFTER** variables to the **Paired Variables** field by clicking (highlighting) these two variables, and then clicking 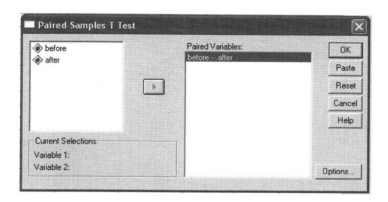 . When the following **Paired-Samples T Test** window opens, click OK to run the T-Test analysis. See Table 5.1 for the results.

5.4.3 SPSS Syntax Method

T-TEST PAIRS=BEFORE AFTER.

5.4.4 SPSS Output

TABLE 5.1

Related T-Test Output

Paired Samples Statistics

		Mean	N	Standard Deviation	Standard Error Mean
Pair 1	BEFORE	177.0000	10	48.3161	15.2789
	AFTER	124.0000	10	43.2563	13.6789

Paired Samples Correlations

		N	Correlation	Sig.
Pair 1	BEFORE & AFTER	10	.745	.013

Paired Samples Test

| | | Paired Differences | | | | | | |
| | | | | 95% Confidence Interval of the Difference | | | | Sig. |
	Mean	Standard Deviation	Standard Error Mean	Lower	Upper	t	df	(2-tailed)
Pair 1 BEFORE - AFTER	53.0000	33.0151	10.4403	29.3824	76.6176	5.076	9	.001

5.4.5   Results and Interpretation

The result from the analysis indicates that there is a significant difference in the amount of food eaten before and after drug X was ingested, t(df = 9) = 5.08, $p < .01$ (see **Paired Samples Test** table). The mean values indicate that significantly less food was consumed after ingestion of drug X (M = 124.00) than before (M = 177.00).

6

One-Way Analysis of Variance, with Post Hoc Comparisons

6.1   Aim

The one-way analysis of variance (ANOVA) is an extension of the independent t-test. It is used when the researcher is interested in whether the means from several (> 2) independent groups differ. For example, if a researcher is interested in investigating whether four ethnic groups differ in their IQ scores, the one-way ANOVA can be used.

6.2   Checklist of Requirements

- In any one analysis, there must be only one independent variable (e.g., ethnicity).
- There should be more than two levels for that independent variable (e.g., Australian, American, Chinese, and African).
- There must be only one dependent variable.

6.3   Assumptions

- The populations from which the samples were taken are normally distributed.
- Homogeneity of variance.
- The observations are all independent of one another.

6.4   Example

A researcher is interested in finding out whether the intensity of electric shock will affect the time required to solve a set of difficult problems. Eighteen subjects are randomly assigned to the three experimental conditions of low shock, medium shock, and high shock. The total time (in minutes) required to solve all the problems is the measure recorded for each subject.

	Shock Intensity				
Low		Medium		High	
s1	15	s7	30	s13	40
s2	10	s8	15	s14	35
s3	25	s9	20	s15	50
s4	15	s10	25	s16	43
s5	20	s11	23	s17	45
s6	18	s12	20	s18	40

6.4.1   Data Entry Format

The data set has been saved under the name: **EX6.SAV**.

Variables	Column(s)	Code
SHOCK	1	1 = low, 2 = medium, 3 = high
TIME	2	Time (in minutes)

6.4.2   Windows Method

1. From the menu bar, click **Analyze**, then **Compare Means**, and then **One-Way ANOVA.** The following **One-Way ANOVA** window will open.

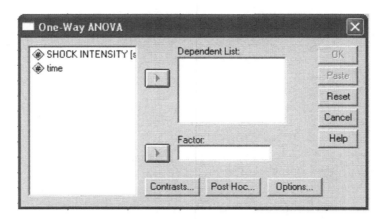

2. Transfer the dependent variable of **TIME** to the **Dependent List** field by clicking (highlighting) the variable and then clicking ▶. Transfer the independent variable of **SHOCK** to the **Factor** field by clicking (highlighting) the variable and then clicking ▶.

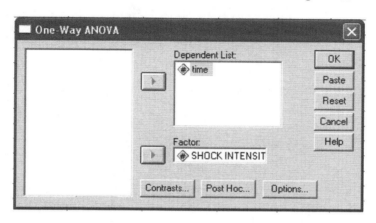

3. Because the one-way ANOVA will only perform an omnibus analysis of the *overall* differences between the three levels (low, medium, and high) of the independent variable of **SHOCK**, it will not analyze the differences between the *specific* shock levels. To obtain multiple comparisons between the three shock levels (low shock vs. medium shock, low shock vs. high shock, medium shock vs. high shock), the researcher needs to perform a *Post Hoc* comparison test. Click **Post Hoc...** to achieve this. When the following **One-Way ANOVA: Post Hoc Multiple Comparisons** window opens, check the **Scheffe** field to run the Scheffé Post Hoc test. Next, click **Continue**.

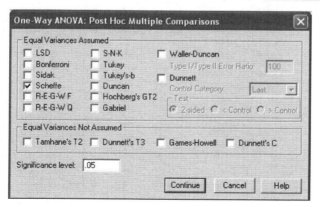

4. When the **One-Way ANOVA** window opens, click Options... to open the **One-Way ANOVA: Options** window. Check the **Descriptive** cell and then click Continue .

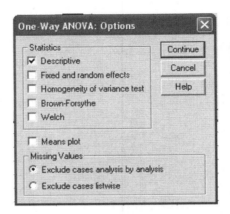

5. When the following **One-Way ANOVA** window opens, run the analysis by clicking OK (see Table 6.1 for the results):

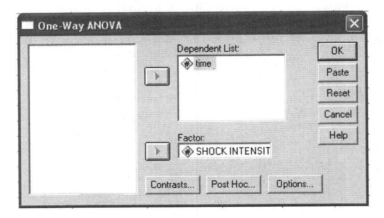

6.4.3   SPSS Syntax Method

ONEWAY TIME BY SHOCK
/STATISTICS=DESCRIPTIVES
/RANGES=SCHEFFE(.05).

6.4.4   SPSS Output

TABLE 6.1

One-Way ANOVA Output

Descriptives

Time	N	Mean	Standard Deviation	Standard Error	95% Confidence Interval for Mean — Lower Bound	95% Confidence Interval for Mean — Upper Bound	Minimum	Maximum
LOW SHOCK	6	17.1667	5.1153	2.0883	11.7985	22.5349	10.00	25.00
MEDIUM SHOCK	6	22.1667	5.1153	2.0883	16.7985	27.5349	15.00	30.00
HIGH SHOCK	6	42.1667	5.1153	2.0883	36.7985	47.5349	35.00	50.00
Total	18	27.1667	12.1086	2.8540	21.1452	33.1881	10.00	50.00

ANOVA

Time	Sum of Squares	df	Mean Square	F	Sig.
Between Groups	2100.000	2	1050.000	40.127	.000
Within Groups	392.500	15	26.167		
Total	2492.500	17			

Post Hoc Tests

Multiple Comparisons
Dependent Variables: TIME
Scheffe

(I) SHOCK INTENSITY	(J) SHOCK INTENSITY	Mean Difference (I – J)	Standard Error	Sig.	95% Confidence Interval — Lower Bound	95% Confidence Interval — Upper Bound
LOW SHOCK	MEDIUM SHOCK	−5.0000	2.95334	.269	−13.0147	3.0147
	HIGH SHOCK	−25.0000*	2.95334	.000	−33.0147	−16.9853
MEDIUM SHOCK	LOW SHOCK	5.0000	2.95334	.269	−3.0147	13.0147
	HIGH SHOCK	−20.0000*	2.95334	.000	−28.0147	−11.9853
HIGH SHOCK	LOW SHOCK	25.0000*	2.95334	.000	16.9853	33.0147
	MEDIUM SHOCK	20.0000*	2.95334	.000	11.9853	28.0147

*The mean difference is significant at the .05 level.

6.4.5   Results and Interpretation

The results from the analysis (Table 6.1) indicate that the intensity of the electric shock has a significant effect on the time taken to solve the problems, $F(2,15) = 40.13$, $p < .001$. The mean values for the three shock levels indicate that as the shock level increased (from low to medium to high), so did the time taken to solve the problems (low: M = 17.17; medium: M = 22.17; high: M = 42.17).

6.4.6   Post Hoc Comparisons

Although the highly significant F-ratio ($p < .001$) indicates that the means of the three shock levels differ significantly, it does not indicate the *location* of this difference. For example, the researcher may want to know whether the overall difference is due primarily to the difference between low shock and high shock levels, between low shock and medium shock levels, or between medium shock and high shock levels. To test for differences between specific shock levels, a number of *post hoc* comparison techniques can be used. For this example, the more conservative **Scheffé** test was used.

In the **Multiple Comparisons** table, in the column labeled **Mean Difference (I – J),** the mean difference values accompanied by asterisks indicate which shock levels differ significantly from each other at the 0.05 level of significance. The results indicate that the high shock level is significantly different from both the low shock and medium shock levels. The low shock level and the medium shock level do not differ significantly. These results show that the overall difference in the time taken to solve complex problems between the three shock-intensity levels is because of the significantly greater amount of time taken by the subjects in the high shock condition.

7

Factorial Analysis of Variance

7.1 Aim

The factorial univariate ANOVA is an extension of the one-way ANOVA in that it involves the analysis of two or more independent variables. It is used in experimental designs in which every level of every factor is paired with every level of every other factor. It allows the researcher to assess the effects of each independent variable separately, as well as the joint effect or inter-action of variables. Factorial designs are labeled either by the *number of factors* involved or in terms of the *number of levels* of each factor. Thus, a factorial design with two independent variables (e.g., gender and ethnicity) and with two levels for each independent variable (male/female; Australian/Chinese) is called either a **two-way factorial** or a **2 × 2 factorial**.

7.2 Checklist of Requirements

- In any one analysis, there must be two or more independent variables (due to the complexity in interpreting higher-order interactions, most factorial designs are limited to three or four independent variables or factors).
- There can be two or more levels for each independent variable.
- There must be only one dependent variable.

7.3   Assumptions

- The populations from which the samples were taken are normally distributed.
- Homogeneity of variance.
- The observations are all independent of one another.

7.4   Example 1 — Two-Way Factorial (2 × 2 Factorial)

A researcher is interested in determining the effects of two learning strategies (A and B) on the memorization of a hard vs. an easy list of syllables. The factorial combination of these two independent variables (2 × 2) yields four experimental conditions: Strategy A–Easy List, Strategy A–Hard List, Strategy B–Easy List, and Strategy B–Hard List. A total of 24 subjects are randomly assigned to the four experimental conditions. The researcher recorded the total number of errors made by each subject.

	Strategy A		Strategy B	
Easy List	s1	6	s13	20
	s2	13	s14	18
	s3	11	s15	14
	s4	8	s16	14
	s5	9	s17	12
	s6	5	s18	16
Hard List	s7	15	s19	16
	s8	17	s20	13
	s9	23	s21	15
	s10	21	s22	20
	s11	22	s23	11
	s12	20	s24	12

7.4.1   Data Entry Format

The data set has been saved under the name: **EX7a.SAV**.

Variables	Column(s)	Code
STRATEGY	1	1 = Strategy A, 2 = Strategy B
LIST	2	1 = Easy, 2 = Hard
ERRORS	3–4	Number of errors made

7.4.2 Windows Method

1. From the menu bar, click **Analyze**, then **General Linear Model**, and then **Univariate.** The following **Univariate** window will open.

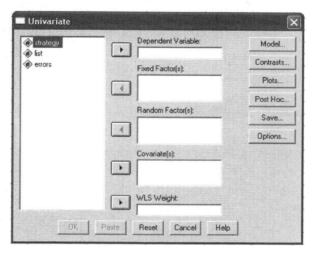

2. Transfer the dependent variable of **ERRORS** to the **Dependent Variable** field by clicking (highlighting) the variable and then clicking

   . Transfer the independent variables of **STRATEGY** and **LIST** to the **Fixed Factor(s)** field by clicking (highlighting) the variables and then clicking .

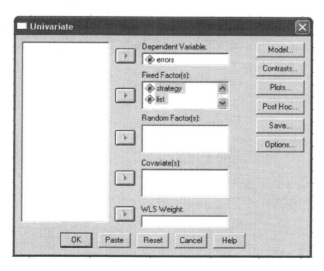

3. Click 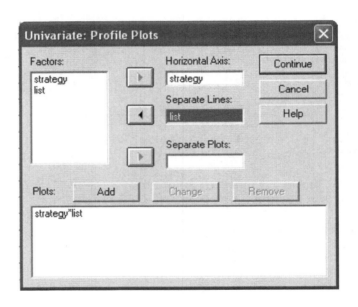 Plots... to plot a graph of the **STRATEGY*LIST** interaction. The following **Univariate: Profile Plots** window will open. Transfer the **STRATEGY** variable to the **Horizontal Axis** field by clicking (highlighting) the variable and then clicking ▶ . Transfer the **LIST** variable to the **Separate Lines** field by clicking (highlighting) the variable and then clicking ▶ . Next, click Add to transfer the **STRATEGY*LIST** interaction to the **Plots** field. When this is done, click Continue .

4. Click Options... in the **Univariate** window to obtain descriptive statistics (Estimated Marginal Means) for the full 2 × 2 **STRATEGY*LIST** interaction. When the **Univariate: Options** window opens, click (highlight) **STRATEGY, LIST,** and **STRATEGY*LIST** in the **Factor(s) and Factor Interactions** field, and then click ▶ to transfer these factors and factor interaction to the **Display Means for** field. Click Continue to return to the **Univariate** window.

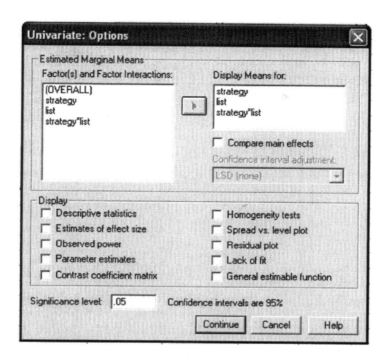

5. When the **Univariate** window opens, click 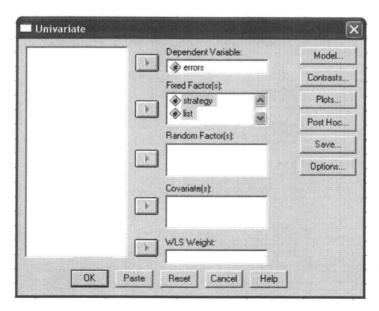 to run the analysis. See Table 7.1 for the results.

7.4.3 SPSS Syntax Method

```
GLM ERRORS BY STRATEGY LIST
/EMMEANS=TABLES(STRATEGY)
/EMMEANS=TABLES(LIST)
/EMMEANS=TABLES(STRATEGY*LIST).
GRAPH
/LINE(MULTIPLE)MEAN(ERRORS) BY STRATEGY BY LIST.
```

7.4.4 SPSS Output

TABLE 7.1

2 × 2 ANOVA Output

Univariate Analysis of Variance Between-Subjects Factors		Value Label	N
strategy	1.00	STRATEGY A	12
	2.00	STRATEGY B	12
list	1.00	EASY	12
	2.00	HARD	12

Tests of Between-Subjects Effects
Dependent Variable: Errors

Source	Type III Sum of Squares	df	Mean Square	F	Sig.
Corrected Model	372.125[a]	3	124.042	13.091	.000
Intercept	5133.375	1	5133.375	541.781	.000
strategy	5.042	1	5.042	.532	.474
list	145.042	1	145.042	15.308	.001
strategy*list	222.042	1	222.042	23.434	.000
Error	189.500	20	9.475		
Total	5695.000	24			
Corrected Total	561.625	23			

[a]R Squared = 0.663 (Adjusted R Squared = 0.612).

Estimated Marginal Means
1. Strategy

Dependent Variable: Errors

strategy	Mean	Standard Error	95% Confidence Interval	
			Lower Bound	Upper Bound
STRATEGY A	14.167	.889	12.313	16.020
STRATEGY B	15.083	.889	13.230	16.937

2. list

Dependent Variable: errors

list	Mean	Standard Error	95% Confidence Interval	
			Lower Bound	Upper Bound
EASY	12.167	.889	10.313	14.020
HARD	17.083	.889	15.230	18.937

3. Strategy*List

Dependent Variable: Errors

strategy	list	Mean	Standard Error	95% Confidence Interval	
				Lower Bound	Upper Bound
STRATEGY A	EASY	8.667	1.257	6.045	11.288
	HARD	19.667	1.257	17.045	22.288
STRATEGY B	EASY	15.667	1.257	13.045	18.288
	HARD	14.500	1.257	11.879	17.121

7.4.5   Results and Interpretation

7.4.5.1   Main Effect

The main effect of **STRATEGY** is not significant, $F(1,20) = 0.53$, $p > .05$ (see Table 7.1). From the estimated marginal means, the difference in the number of errors made by the **strategy A** group (M = 14.167) is not significantly different from the number of errors made by the **strategy B** group (M = 15.083) (collapsing across the two **LIST** levels).

The main effect of **LIST** is significant, $F(1,20) = 15.31$, $p < .05$. From the Estimated Marginal Means, it can be seen that the subjects made significantly more errors in the **hard list** (M = 17.08) than in the **easy list** (M = 12.16) (collapsing across the two **STRATEGY** levels).

7.4.5.2   Interaction Effect

The **STRATEGY*LIST** interaction is significant, $F(1,20) = 23.43$, $p < .001$. To interpret the interaction, the task is made easier by graphing the **STRATEGY*LIST** estimated marginal means from Table 7.1, as shown in Figure 7.1.

From Figure 7.1, it can be seen that the effect of learning strategy on the number of errors made is dependent on the difficulty of the list learned. Under strategy A, subjects made more errors in the hard list than in the easy list, but under strategy B, the effect is opposite, with subjects making more errors in the easy list than in the hard list.

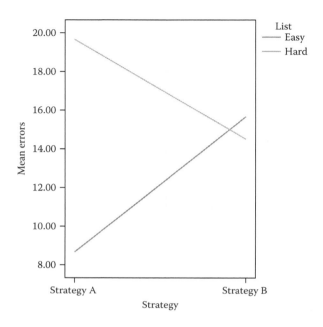

FIGURE 7.1

7.4.6   Post Hoc Test for Simple Effects

The significant interaction effect indicates that the effect of one independent variable on the dependent variable is dependent on the second independent variable, i.e., the four experimental conditions (Strategy A–Easy List, Strategy A–Hard List, Strategy B–Easy List, and Strategy B–Hard List) differ significantly in affecting the number of errors made. However, the interaction effect does not indicate where the differences are, i.e., between which experimental conditions. To identify specific differences, **Post Hoc** comparisons can be used to "tease apart" the interaction. This is equivalent to the test for simple effects, i.e., the effect of one factor (IV_1) at one level of the other factor (IV_2).

Unfortunately, in a factorial design such as this, post hoc comparisons between the four experimental conditions cannot be directly conducted within the GLM Univariate program. Rather, post hoc comparisons can only be executed through the **ONEWAY ANOVA** analysis. This procedure requires some data manipulation to convert the four experimental conditions into four levels of the same grouping variable. These four levels can then be compared using the **Scheffé** post hoc test via the **ONEWAY ANOVA** analysis.

7.4.7 Data Transformation

7.4.7.1 *Windows Method*

1. The first step is to create a new grouping variable called **GROUP** that contains the four levels generated by the **STRATEGY*LIST** interaction. From the menu bar, click **Transform** and then **Compute**. The following **Compute Variable** window will open:

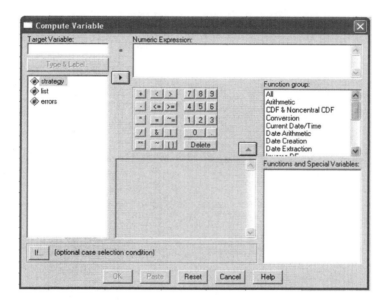

2. Click If... to open the **Compute Variable: If Cases** window. Ensure that the **Include if case satisfies condition** field is checked. Using the data entry format codes presented in Subsection 7.4.1, create the first level (**Strategy A–Easy List**) by typing **strategy=1 and list=1** in the field. Click Continue.

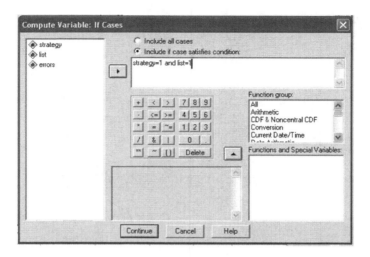

3. Because the **Strategy A–Easy List** level is the first of four levels within a new grouping variable called **GROUP**, this level will be coded **1** within the **GROUP** variable. When the following **Compute Variable** window opens, type **GROUP** in the **Target Variable** field and **1** in the **Numeric Expression** field. Click ▭ to create the first level of **Strategy A–Easy List** (coded **GROUP=1**) within the new grouping variable of **GROUP**.

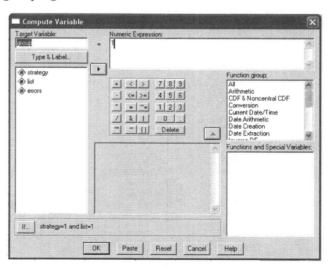

4. Repeat steps 1 to 3 to create the other three levels: **Strategy A–Hard List** (coded **GROUP=2**), **Strategy B–Easy List** (coded **GROUP=3**), and **Strategy B–Hard List** (coded **GROUP=4**). For example, to create the second level of **Strategy A–Hard List,** open the **Compute**

Variable: If Cases window. Type **strategy=1 and list=2** in the **Include if case satisfies condition** field. Click Continue .

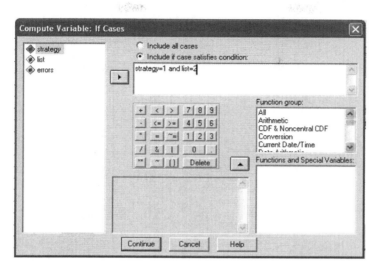

When the following **Compute Variable** window opens, type **GROUP** in the **Target Variable** field and **2** in the **Numeric Expression** field. Click OK to create the second level of **Strategy A–Hard List** (coded **GROUP=2**) within the new grouping variable of **GROUP**.

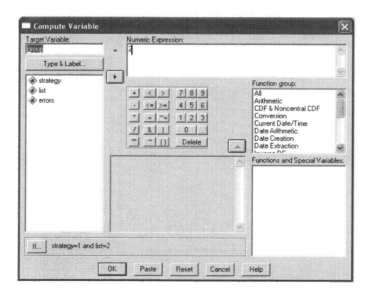

5. To aid interpretation of the obtained results, **Value Labels** in the data set should be activated and labels attached to the numerical codes for the four levels. To do this, open the data set, and under **Variable View**, click the **Values** field for the **GROUP** variable. Type in the value labels as indicated in the **Value Labels** window in the following:

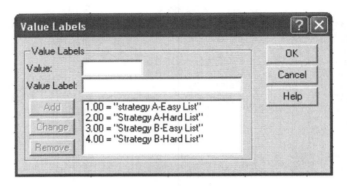

7.4.7.2   *Post Hoc Comparisons: Windows Method*

Once the four levels have been created, **Scheffé** Post Hoc comparisons can be carried out to test for differences (simple effects) between these four levels.

1. From the menu bar, click **Analyze**, then **Compare Means**, and then **One-Way ANOVA**. The following **One-Way ANOVA** window will open. Note that the listing of variables now includes the newly created grouping variable of **GROUP** that contains the four levels of **Strategy A–Easy List**, **Strategy A–Hard List**, **Strategy B–Easy List**, and **Strategy B–Hard List**.

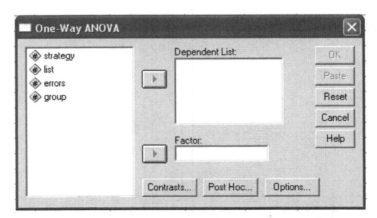

2. Transfer the dependent variable of **ERRORS** to the **Dependent List** field by clicking (highlighting) the variable and then clicking [▶]. Transfer the independent variable of **GROUP** to the **Factor** field by clicking (highlighting) the variable and then clicking [▶]. Click [Post Hoc...] to execute a post hoc comparison between the four levels (**Strategy A–Easy List, Strategy A Hard List, Strategy B–Easy List,** and **Strategy B–Hard List**).

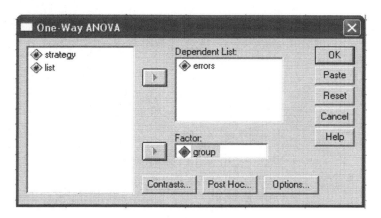

3. When the following **One-Way ANOVA: Post Hoc Multiple Comparisons** window opens, check the **Scheffe** field to run the Scheffé Post Hoc test. Next, click [Continue].

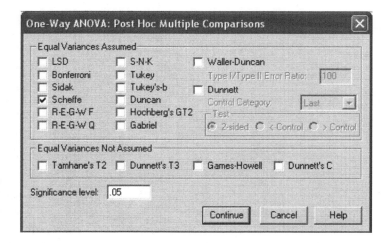

4. When the **One-Way ANOVA** window opens, run the analysis by clicking 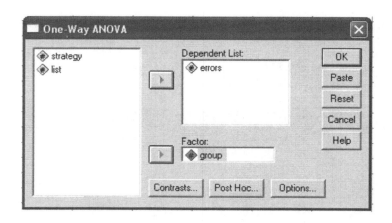. See Table 7.2 for the results.

7.4.7.3   *Post Hoc Comparisons: SPSS Syntax Method*
IF (STRATEGY EQ 1 AND LIST EQ 1) GROUP=1.
IF (STRATEGY EQ 1 AND LIST EQ 2) GROUP=2.
IF (STRATEGY EQ 2 AND LIST EQ 1) GROUP=3.
IF (STRATEGY EQ 2 AND LIST EQ 2) GROUP=4.

VALUE LABELS GROUP   1 'STRATEGY A–Easy LIST'
                            2 'STRATEGY A–Hard LIST'
                            3 'STRATEGY B–Easy LIST'
                            4 'STRATEGY B–Hard LIST'.
    ONEWAY ERRORS BY GROUP(1,4)
    /RANGES=SCHEFFE (.05).

7.4.8 SPSS Output

TABLE 7.2

Scheffé Post Hoc Comparisons Output

Multiple Comparisons

Dependent Variable: Errors
Scheffe

(I) group	(J) group	Mean Difference (I – J)	Standard Error	Sig.	95% Confidence Interval Lower Bound	Upper Bound
Strategy A–Easy List	Strategy A–Hard List	–11.00000*	1.77717	.000	–16.4182	–5.5818
	Strategy B–Easy List	–7.00000*	1.77717	.008	–12.4182	–1.5818
	Strategy B–Hard List	–5.83333*	1.77717	.032	–11.2516	–0.4151
Strategy A–Hard List	Strategy A–Easy List	11.00000*	1.77717	.000	5.5818	16.4182
	Strategy B–Easy List	4.00000	1.77717	.201	–1.4182	9.4182
	Strategy B–Hard List	5.16667	1.77717	.065	–0.2516	10.5849
Strategy B–Easy List	strategy A–Easy List	7.00000*	1.77717	.008	1.5818	12.4182
	Strategy A–Hard List	–4.00000	1.77717	.201	–9.4182	1.4182
	Strategy B–Hard List	1.16667	1.77717	.933	–4.2516	6.5849
Strategy B–Hard List	strategy A–Easy List	5.83333*	1.77717	.032	0.4151	11.2516
	Strategy A–Hard List	–5.16667	1.77717	.065	–10.5849	0.2516
	Strategy B–Easy List	–1.16667	1.77717	.933	–6.5849	4.2516

*The mean difference is significant at the .05 level.

7.4.9 Results and Interpretation

Results from Scheffé comparisons (see Table 7.2) indicate that the significant **STRATEGY*LIST** interaction is due primarily to the subjects making significantly less errors in the Strategy A–Easy List condition (M = 8.66) than in the Strategy B–Easy List (M = 15.66), Strategy A–Hard List (M = 19.66), and Strategy B–Hard List (M = 14.50) conditions. No other conditions are significantly different from each other.

7.5 Example 2 — Three-Way Factorial (2 × 2 × 2 Factorial)

The previous example can be extended to a three-way factorial design. Assume that the researcher, in addition to determining the effects of the two types of learning strategies (A and B) on the memorization of easy-vs.-difficult list, is also concerned with determining the effects of high-vs.-low-intensity shock on learning in the conditions mentioned earlier. The design includes shock level as an additional factor, making the entire experiment a three-factor design. The researcher records the total number of errors made by each subject. The three-way factorial combination of the three independent variables yields the following eight experimental groups, with six subjects per group.

| | Strategy A | | | | Strategy B | | | |
	Easy List		Hard List		Easy List		Hard List	
Low	s1	17	s7	25	s13	13	s19	27
Shock	s2	10	s8	18	s14	18	s20	29
	s3	11	s9	19	s15	10	s21	31
	s4	9	s10	19	s16	16	s22	36
	s5	10	s11	16	s17	22	s23	40
	s6	6	s12	13	s18	18	s24	28
High	s25	16	s31	26	s37	13	s43	41
Shock	s26	8	s32	18	s38	19	s44	34
	s27	5	s33	12	s39	14	s45	40
	s28	7	s34	20	s40	16	s46	41
	s29	8	s35	15	s41	23	s47	35
	s30	9	s36	19	s42	20	s48	33

7.5.1   Data Entry Format

The data set has been saved under the name: **EX7b.SAV**.

Variables	Column(s)	Code
STRATEGY	1	1 = Strategy A, 2 = Strategy B
LIST	2	1 = Easy, 2 = Hard
SHOCK	3	1 = Low, 2 = High
ERRORS	4	Number of errors made

7.5.2   Windows Method

1. From the menu bar, click **Analyze**, then **General Linear Model**, and then **Univariate.** The following **Univariate** window will open.

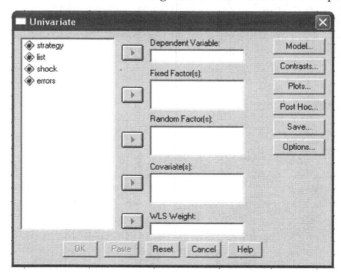

2. Transfer the dependent variable of **ERRORS** to the **Dependent Variable** field by clicking (highlighting) the variable and then clicking

   [▶] . Transfer the independent variables of **STRATEGY, LIST,** and **SHOCK** to the **Fixed Factor(s)** field by clicking (highlighting) the

   variables and then clicking [▶] .

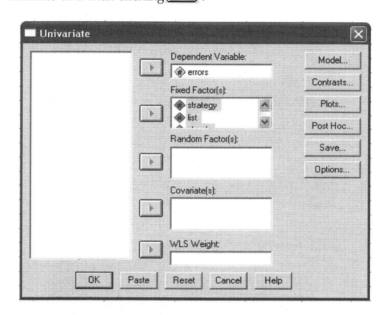

3. Click [Options...] to obtain means for the **STRATEGY, LIST,** and **SHOCK** factors, and the **STRATEGY*LIST, STRATEGY*SHOCK, LIST*SHOCK,** and **STRATEGY*LIST*SHOCK** interactions. When the **Univariate: Options** window opens, click (highlight) **STRATEGY, LIST, SHOCK, STRATEGY*LIST, STRATEGY*SHOCK, LIST*SHOCK,** and **STRATEGY*LIST*SHOCK** in the **Factor(s) and**

   Factor Interactions field, and then click [▶] to transfer these factors and factor interactions to the **Display Means for** field. Click

   [Continue] to return to the **Univariate** window.

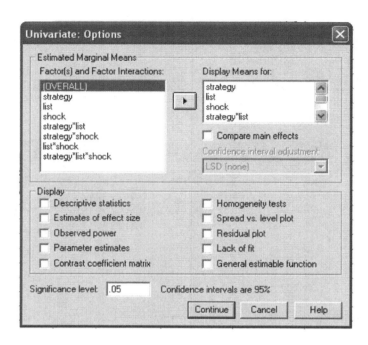

4. When the **Univariate** window opens, click 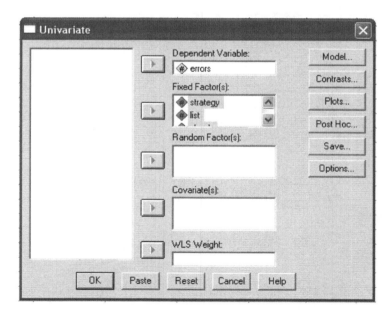 to plot graphs of the three two-way interactions (**STRATEGY*LIST**, **STRATEGY*SHOCK**, and **LIST*SHOCK**).

5. When the following **Univariate: Profile Plots** window opens, transfer the **STRATEGY** variable to the **Horizontal Axis** field by clicking

   (highlighting) the variable and then clicking ▶ . Transfer the **LIST** variable to the **Separate Lines** field by clicking (highlighting)

   the variable and then clicking ▶ . Next, click Add to transfer the **STRATEGY*LIST** interaction to the **Plots** field. Repeat this procedure to add the **STRATEGY*SHOCK** and **LIST*SHOCK**

   interactions to the **Plots** field. When this is done, click Continue .

6. To plot the three-way **STRATEGY*LIST*SHOCK** interaction, some data transformation must be carried out. Suppose the researcher decides to plot the variables of **STRATEGY** and **LIST** (i.e., **STRATEGY*LIST** interaction) against the variable of **SHOCK**. The first step is to create a new grouping variable called **GROUP** that contains the four levels generated by the **STRATEGY*LIST** interaction (Strategy A–Easy List, Strategy A–Hard List, Strategy B–Easy List, and Strategy B–Hard List). To do this, follow step 1 through to step 5 in Subsection 7.4.7.1.

7. When the data transformation procedure has been completed, the newly created grouping variable of **GROUP** (containing the four levels of Strategy A–Easy List, Strategy A–Hard List, Strategy B–Easy List, and Strategy B–Hard List) will be added to the data set. Click

Plots... in the **Univariate** window to open the **Univariate: Profile Plots** window. Transfer the **SHOCK** variable to the **Horizontal Axis** field by clicking (highlighting) the variable and then clicking

▶ . Transfer the newly created **GROUP** variable to the **Separate Lines** field by clicking (highlighting) the variable and then clicking

▶ . Next, click Add to transfer the **SHOCK*GROUP** interaction to the **Plots** field. When this is done, click Continue .

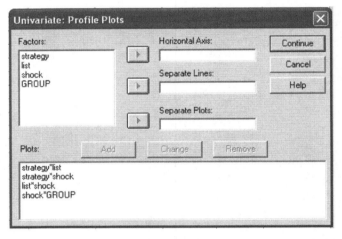

8. When the **Univariate** window opens, run the analysis by clicking

OK . See Table 7.3 for the results.

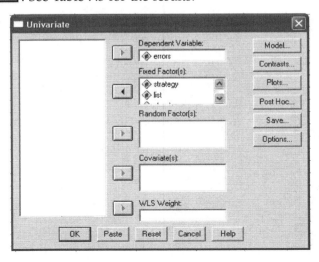

7.5.3   SPSS Syntax Method

GLM ERRORS BY STRATEGY LIST SHOCK
/EMMEANS=TABLES(STRATEGY)
/EMMEANS=TABLES(LIST)
/EMMEANS=TABLES(SHOCK)
/EMMEANS=TABLES(STRATEGY*LIST)
/EMMEANS=TABLES(STRATEGY*SHOCK)
/EMMEANS=TABLES(LIST*SHOCK).
/EMMEANS=TABLES(STRATEGY*LIST*SHOCK

IF (STRATEGY EQ 1 AND LIST EQ 1) GROUP=1.
IF (STRATEGY EQ 1 AND LIST EQ 2) GROUP=2.
IF (STRATEGY EQ 2 AND LIST EQ 1) GROUP=3.
IF (STRATEGY EQ 2 AND LIST EQ 2) GROUP=4.
VALUE LABELS GROUP   1 'STRATEGY A–Easy LIST'

                       2 'STRATEGY A–Hard LIST'

                       3 'STRATEGY B–Easy LIST'

                       4 'STRATEGY B–Hard LIST'.
GRAPH
/LINE(MULTIPLE)MEAN(ERRORS) BY STRATEGY BY LIST.
GRAPH
/LINE(MULTIPLE)MEAN(ERRORS) BY STRATEGY BY SHOCK.
GRAPH
/LINE(MULTIPLE)MEAN(ERRORS) BY LIST BY SHOCK.
GRAPH
/LINE(MULTIPLE)MEAN(ERRORS) BY SHOCK BY GROUP.

7.5.4   SPSS Output

TABLE 7.3

General Linear Model
$2 \times 2 \times 2$ ANOVA Output

| | | Between-Subjects Factors | |
		Value Label	N
Strategy	1.00	STRATEGY A	24
	2.00	STRATEGY B	24
List	1.00	EASY	24
	2.00	HARD	24
Shock	1.00	LOW	24
	2.00	HIGH	24

Tests of Between-Subjects Effects

Dependent Variable: errors

Source	Type III Sum of Squares	df	Mean Square	F	Sig.
Corrected Model	4090.479[a]	7	584.354	33.802	.000
Intercept	18921.021	1	18921.021	1094.491	.000
strategy	1645.021	1	1645.021	95.157	.000
list	2093.521	1	2093.521	121.100	.000
shock	20.021	1	20.021	1.158	.288
strategy * list	247.521	1	247.521	14.318	.001
strategy * shock	54.187	1	54.187	3.134	.084
list * shock	25.521	1	25.521	1.476	.231
strategy * list * shock	4.688	1	4.688	.271	.605
Error	691.500	40	17.288		
Total	23703.000	48			
Corrected Total	4781.979	47			

[a]R Squared = .855 (Adjusted R Squared = .830)

Estimated Marginal Means

1. Strategy

Dependent Variable: Errors

Strategy	Mean	Standard Error	95% Confidence Interval	
			Lower Bound	Upper Bound
STRATEGY	14.000	.849	12.285	15.715
STRATEGY B	25.708	.849	23.993	27.424

2. List

Dependent Variable: Errors

List	Mean	Standard Error	95% Confidence Interval	
			Lower Bound	Upper Bound
EASY	13.250	.849	11.535	14.965
HARD	26.458	.849	24.743	28.174

3. Shock

Dependent Variable: Errors

Shock	Mean	Standard Error	95% Confidence Interval	
			Lower Bound	Upper Bound
LOW	19.208	.849	17.493	20.924
HIGH	20.500	.849	18.785	22.215

4. Strategy * list

Dependent Variable: Errors

Strategy	List	Mean	Standard Error	95% Confidence Interval	
				Lower Bound	Upper Bound
STRATEGY A	EASY	9.667	1.200	7.241	12.092
	HARD	18.333	1.200	15.908	20.759
STRATEGY B	EASY	16.833	1.200	14.408	19.259
	HARD	34.583	1.200	32.158	37.009

5. Strategy * shock

Dependent Variable: Errors

Strategy	Shock	Mean	Standard Error	95% Confidence Interval	
				Lower Bound	Upper Bound
STRATEGY A	LOW	14.417	1.200	11.991	16.842
	HIGH	13.583	1.200	11.158	16.009
STRATEGY B	LOW	24.000	1.200	21.574	26.426
	HIGH	27.417	1.200	24.991	29.842

6. List* shock

Dependent Variable: Errors

List	Shock	Mean	Standard Error	95% Confidence Interval	
				Lower Bound	Upper Bound
EASY	LOW	13.333	1.200	10.908	15.759
	HIGH	13.167	1.200	10.741	15.592
HARD	LOW	25.083	1.200	22.658	27.509
	HIGH	27.833	1.200	25.408	30.259

7. Strategy * list * shock

Dependent Variable: Errors

Strategy	List	Shock	Mean	Standard Error	95% Confidence Interval	
					Lower Bound	Upper Bound
STRATEGY A	EASY	LOW	10.500	1.697	7.069	13.931
		HIGH	8.833	1.697	5.403	12.264
	HARD	LOW	18.333	1.697	14.903	21.764
		HIGH	18.333	1.697	14.903	21.764
STRATEGY B	EASY	LOW	16.167	1.697	12.736	19.597
		HIGH	17.500	1.697	14.069	20.931
	HARD	LOW	31.833	1.697	28.403	35.264
		HIGH	37.333	1.697	33.903	40.764

7.5.5   Results and Interpretation

7.5.5.1   Main Effects

The main effect of **STRATEGY** is significant, $F(1,40) = 95.16$, $p < .001$ (see Table 7.3). From the estimated marginal means, subjects made significantly more errors under **strategy B** (M = 25.71) than under **strategy A** (M = 14.00) (collapsing across the **LIST** and **SHOCK** factors).

The main effect of **LIST** is significant, $F(1,40) = 121.10$, $p < .001$. Subjects made significantly more errors in the **hard list** condition (M = 26.45) than in the **easy list** condition (M = 13.25) (collapsing across the **STRATEGY** and **SHOCK** factors).

The main effect for **SHOCK** is not significant, $F(1,40) = 1.16$, $p > .05$. The difference in the number of errors made under the low-shock condition (M = 19.21) is not significantly different from the number of errors made under the high-shock condition (M = 20.50) (collapsing across the **LIST** and **STRATEGY** factors).

7.5.5.2   Two-Way Interactions

*7.5.5.2.1   Strategy*List Interaction*

The **STRATEGY*LIST** interaction is significant, $F(1,40) = 14.32$, $p < .01$. Interpretation of this interaction can be facilitated by graphing the **STRATEGY*LIST** estimated marginal means from Table 7.3, as shown in Figure 7.2.

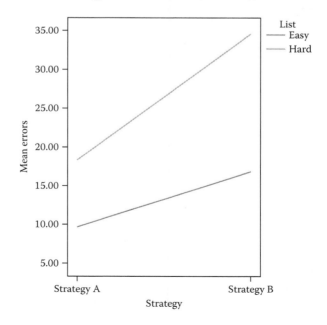

FIGURE 7.2

The significant interaction effect indicates that the effect of learning strategy on the number of errors made is dependent on the difficulty of the list learned. Although the number of errors made increased from Strategy A to Strategy B when learning either hard or easy list, the increase is more pronounced when learning the hard list than the easy list.

7.5.5.2.2 Post Hoc Comparisons

Post hoc comparisons can now be carried out to clarify the preceding interaction, i.e., to locate differences between the four experimental conditions. The **Windows** method for carrying out these post hoc comparisons is identical to that presented in Subsection 7.4.7.2, from step 1 to step 4. The **SPSS syntax method** used for carrying out these comparisons is identical to that presented in Subsection 7.4.7.3. See Table 7.4 for the results.

7.5.5.2.3 SPSS Output

TABLE 7.4

Scheffé Post Hoc Comparisons

Multiple Comparisons

Dependent Variable: Errors
Scheffe

(I) GROUP	(J) GROUP	Mean Difference (I – J)	Standard Error	Sig.	Lower Bound	Upper Bound
STRATEGY A– EASY LIST	STRATEGY A–HARD LIST	−8.66667*	1.73633	.000	−13.7138	−3.6195
	STRATEGY B–EASY LIST	−7.16667*	1.73633	.002	−12.2138	−2.1195
	STRATEGY B–HARD LIST	−24.91667*	1.73633	.000	−29.9638	−19.8695
STRATEGY A– HARD LIST	STRATEGY A–EASY LIST	8.66667*	1.73633	.000	3.6195	13.7138
	STRATEGY B–EASY LIST	1.50000	1.73633	.862	−3.5471	6.5471
	STRATEGY B–HARD LIST	−16.25000*	1.73633	.000	−21.2971	−11.2029
STRATEGY B– EASY LIST	STRATEGY A–EASY LIST	7.16667*	1.73633	.002	2.1195	12.2138
	STRATEGY A–HARD LIST	−1.50000	1.73633	.862	−6.5471	3.5471
	STRATEGY B–HARD LIST	−17.75000*	1.73633	.000	−22.7971	−12.7029
STRATEGY B– HARD LIST	STRATEGY A–EASY LIST	24.91667*	1.73633	.000	19.8695	29.9638
	STRATEGY A–HARD LIST	16.25000*	1.73633	.000	11.2029	21.2971
	STRATEGY B–EASY LIST	17.75000*	1.73633	.000	12.7029	22.7971

*The mean difference is significant at the .05 level.

7.5.5.2.4 Results and Interpretation

Results from the post hoc comparisons (Table 7.4) indicate that the significant **STRATEGY*LIST** interaction is due primarily to subjects making significantly less errors in the Strategy A–Easy List condition than in the other three experimental conditions (A–Hard, B–Easy, and B–Hard), and to subjects making significantly more errors in the Strategy B–Hard List condition than in the other three experimental conditions (B–Easy, A–Easy, and A–Hard).

7.5.5.2.5   *STRATEGY*SHOCK Interaction*

The **STRATEGY*SHOCK** interaction is not significant, $F(1,40)$ = 3.13, $p >$ .05. Interpretation of this interaction can be facilitated by graphing the **STRATEGY*SHOCK** estimated marginal means from Table 7.3, as shown in Figure 7.3.

As the interaction is not significant, the result can be interpreted in terms of the significant main effect for **STRATEGY**. That is, the effect of **STRATEGY** on the number of errors made is not dependent on the levels of **SHOCK**, such that regardless of **SHOCK** level, subjects made significantly more errors under Strategy B (M = 25.71) than under Strategy A (M = 14.00).

7.5.5.2.6   *LIST*SHOCK Interaction*

The **LIST*SHOCK** interaction is not significant, $F(1,40)$ = 1.47, $p >$ .05. Interpretation of this interaction can be facilitated by graphing the **LIST*SHOCK** estimated marginal means from Table 7.3, as shown in Figure 7.4

The effect of **LIST** on the number of errors made is not dependent on the levels of **SHOCK**, such that regardless of **SHOCK** level, subjects made more errors on the hard list (M = 26.45) than on the easy list (M = 13.25).

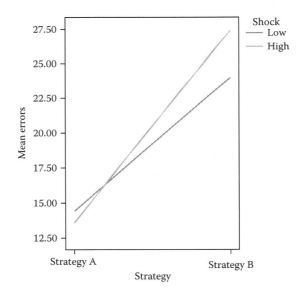

FIGURE 7.3

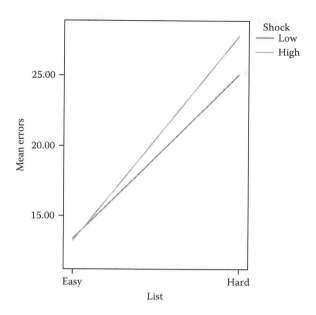

FIGURE 7.4

7.5.5.3   Three-Way Interaction

7.5.5.3.1   STRATEGY*LIST*SHOCK Interaction

The three-way interaction between **STRATEGY, LIST,** and **SHOCK** is not significant, $F(1,40) = 0.27$, $p > .05$. Interpretation of this interaction can be facilitated by graphing the **STRATEGY*LIST*SHOCK** estimated marginal means from Table 7.3, as shown in Figure 7.5.

As the three-way interaction is not significant, it is legitimate to interpret the significant main effects of **LIST** and **STRATEGY.** For example, the results indicate that more errors were made on the hard list (M = 26.45) than on the easy list (M = 13.25), and under Strategy B (M = 25.71) than under Strategy A (M = 14.00), regardless of shock level.

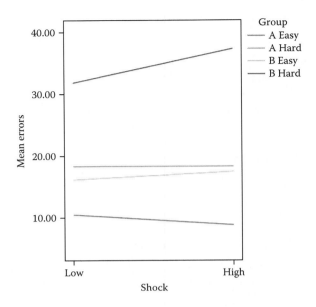

FIGURE 7.5

8

General Linear Model (GLM) Multivariate Analysis

8.1 Aim

In experiments involving multiple independent variables and one dependent variable, the general linear model (GLM) *univariate analysis of variance* is usually used to answer questions about the effects of the independent variables on the dependent variable. The last example in Chapter 7 examined the effects of three independent variables (type of learning strategy, shock level, and difficulty of material) on the dependent variable of the number of errors made by each subject in learning one list of material. As the experiment involved only one dependent variable, the GLM univariate (2 × 2 × 2) analysis of variance was the appropriate test to use. However, if the experiment had required each subject to learn four different lists of material (instead of one list), the GLM univariate analysis would no longer be appropriate. This is because the dependent variable is no longer a single measure but four different scores obtained for each subject. Although the GLM univariate analysis of variance can be conducted separately for each of the four dependent variables, the *GLM multivariate analysis of variance* is more appropriate. Unlike univariate tests, GLM multivariate analysis takes into account the interrelation among dependent variables and analyzes the variables simultaneously.

8.2 Checklist of Requirements

- Depending on the research question and hypotheses to be tested, the experiment can include or exclude "classification" on independent variables.

- When independent variables are included, there can be two or more levels for each independent variable.
- There should be two or more dependent variables.

8.3   Assumptions

- The observations must be independent (i.e., responses among groups of respondents should not be correlated).
- The variance–covariance matrices must be equal for all treatment groups.
- The set of dependent variables must follow a multivariate normal distribution (i.e., any linear combination of the dependent variables must follow a normal distribution).

8.4   Example 1 — GLM Multivariate Analysis: One-Sample Test

The one-sample GLM model incorporating multiple dependent variables is an extension of the one-sample t-test and is the simplest example of GLM. Where the one-sample t-test is used to test the hypothesis that the sample does not differ from a population with a known mean, the one-sample GLM tests the hypothesis that several observed means do not differ from a set of constants. That is, it tests the hypothesis that a set of means is equal to zero.

Suppose that a sports psychologist has recorded the following running times (in seconds) from five men in four different events: 50-yd dash, 100-yd dash, 200-yd dash, and 300-yd dash. The hypothesis to be tested is that the observed sample comes from a population with specified values for the means of the running events. For illustrative purposes, the standard values are taken to be 6 sec for the 50-yd dash, 12 sec for the 100-yd dash, 25 sec for the 200-yd dash, and 40 sec for the 300-yd dash. Because GLM automatically tests the hypothesis that a set of means is equal to zero, the normative values must be subtracted from the observed scores, and the hypothesis that the differences are zero is tested. The running times for the four events are presented in the following:

	Running Events			
	50 yd	100 yd	200 yd	300 yd
s1	9	20	42	67
s2	9	19	45	66
s3	8	17	38	62
s4	9	18	46	64
s5	8	22	39	59

8.4.1   Data Entry Format

The data set has been saved under the name: **EX8a.SAV**.

Variables	Column(s)	Code
T1	1	Running speed in sec
T2	2	Running speed in sec
T3	3	Running speed in sec
T4	4	Running speed in sec

8.4.2   Windows Method

1. To test the hypothesis that a set of means is equal to zero, the normative values must first be subtracted from the observed scores. From the menu bar, click **Transform**, and then **Compute**. The following **Compute Variable** window will open:

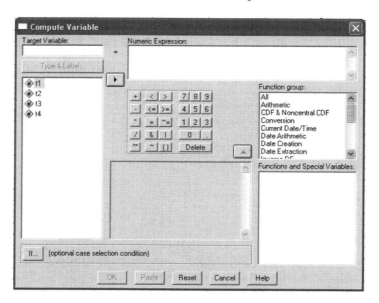

2. To create the first variable representing the difference between the recorded time for the 50-yd dash and its normative time of 6 sec, type the name of the variable **Fifty** in the **Target Variable** field. In the **Numeric Expression** field, type the expression **T1-6**. Next, click

   OK . This will create a new variable called **Fifty**, which represents the difference between the observed score (T1: time in sec to run 50 yd dash) and its normative value (6 sec).

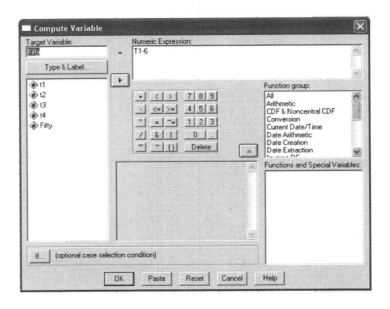

3. To create the other three variables representing the 100-yd dash (**ONEHUND**), the 200-yd dash (**TWOHUND**), and the 300-yd dash (**THREEHUN**), follow the steps presented in step 2. Note that for each of the three computed variables (**ONEHUND, TWOHUND, THREEHUN**), its normative value has been subtracted from its observed score (i.e., T2-12, T3-25, T4-40). Successful completion of this procedure will create the four new variables of **FIFTY, ONEHUND, TWOHUND,** and **THREEHUN.**

4. The next step is to run the one-sample GLM to test the hypothesis that the means of the four computed variables (**FIFTY, ONEHUND, TWOHUND,** and **THREEHUN**) do not differ from a set of constants. From the menu bar, click **Analyze**, then **General Linear Model**, and then **Multivariate**. The following **Multivariate** window will open:

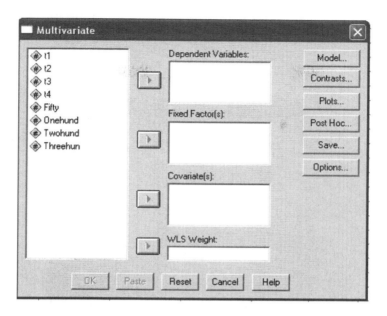

5. Transfer the four variables of **FIFTY, ONEHUND, TWOHUND,** and **THREEHUN** to the **Dependent Variables** field by clicking (high-lighting) these four variables and then clicking 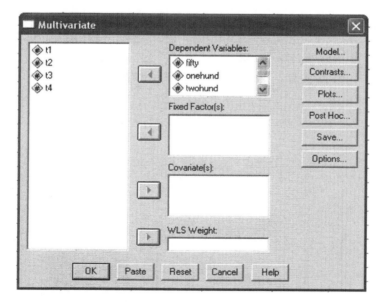.

6. To obtain descriptive statistics for these four variables, click Options... . This will open the **Multivariate: Options** window. Check the **Descriptive statistics** cell and then click Continue .

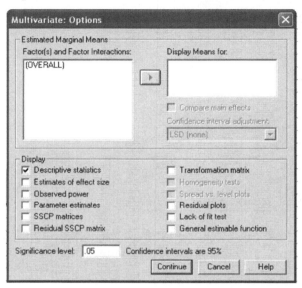

7. This will open the **Multivariate** window. Click OK to complete the analysis. See Table 8.1 for the results.

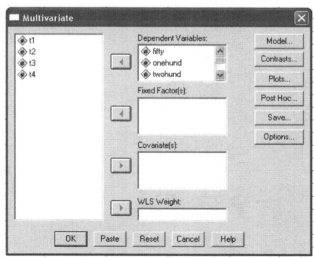

8.4.3   SPSS Syntax Method

COMPUTE FIFTY=(T1-6).
COMPUTE ONEHUND=(T2-12).
COMPUTE TWOHUND=(T3-25).
COMPUTE THREEHUN=(T4-40).

GLM FIFTY ONEHUND TWOHUND THREEHUN
/PRINT=DESCRIPTIVES.

8.4.4   SPSS Output

TABLE 8.1

One-Sample GLM Output

| | Descriptive Statistics | | |
	Mean	Standard Deviation	N
FIFTY	2.6000	.5477	5
ONEHUND	7.2000	1.9235	5
TWOHUND	17.0000	3.5355	5
THREEHUN	23.6000	3.2094	5

| | | | | Multivariate Tests[b] | | |
Effect		Value	F	Hypothesis df	Error df	Sig.
Intercept	Pillai's Trace	.999	232.563[a]	4.000	1.000	.049
	Wilks' Lambda	.001	232.563[a]	4.000	1.000	.049
	Hotelling's Trace	930.250	232.563[a]	4.000	1.000	.049
	Roy's Largest Root	930.250	232.563[a]	4.000	1.000	.049

[a]Exact statistic.
[b]Design: Intercept.

		Tests of between-Subjects Effects				
Source	Dependent Variable	Type III Sum of Squares	df	Mean Square	F	Sig.
Corrected Model	FIFTY	.000[a]	0	.	.	.
	ONEHUND	.000[a]	0	.	.	.
	TWOHUND	.000[a]	0	.	.	.
	THREEHUN	.000[a]	0	.	.	.
Intercept	FIFTY	33.800	1	33.800	112.667	.000
	ONEHUND	259.200	1	259.200	70.054	.001
	TWOHUND	1445.000	1	1445.000	115.600	.000
	THREEHUN	2784.800	1	2784.800	270.369	.000
Error	FIFTY	1.200	4	.300		
	ONEHUND	14.800	4	3.700		
	TWOHUND	50.000	4	12.500		
	THREEHUN	41.200	4	10.300		
Total	FIFTY	35.000	5			
	ONEHUND	274.000	5			
	TWOHUND	1495.000	5			
	THREEHUN	2826.000	5			
Corrected Total	FIFTY	1.200	4			
	ONEHUND	14.800	4			
	TWOHUND	50.000	4			
	THREEHUN	41.200	4			

[a]R Squared = .000 (Adjusted R Squared = .000).

8.4.5   Results and Interpretation

The **Descriptive Statistics** table presents means and standard deviations for the four running events after their normative values have been subtracted. The sample exceeds (i.e., is poorer than) the norm for all four running events.

The **Multivariate Tests** table tests the hypothesis that the four sample means do not differ from the specified set of constants. Of the three multivariate tests (Pillai's, Hotelling's, Wilks', and Roy's), Pillai's trace is the most powerful (i.e., the ability to detect differences if they exist) and the most robust (i.e., the significance level based on it is reasonably correct even when the assumptions are violated). Because the observed significance level is small ($p < .05$), the null hypothesis that the sample means do not differ from the specified constants is rejected, multivariate Pillai $F(1,4) = 232.56$, $p < .05$.

When the null hypothesis of no difference is rejected, it is often informative to examine the univariate **Tests of Between-Subjects Effects** to identify which variables yielded significant differences. The univariate test results (**Intercept**) show that the means for all running events differ significantly from their specified standards ($p < .01$).

8.5 Example 2 — GLM Multivariate Analysis: Two-Sample Test

Following from the previous example, suppose that the researcher also recorded running times for the four events for five women. The running times for the two samples for the four events are presented in the following. The hypothesis that men and women do not differ on the four running events will be tested.

Running Events

Men	50 yd	100 yd	200 yd	300 yd
s1	9	20	42	67
s2	9	19	45	66
s3	8	17	38	62
s4	9	18	46	64
s5	8	22	39	59

Women	50 yd	100 yd	200 yd	300 yd
s1	15	30	41	77
s2	14	29	44	76
s3	13	27	39	72
s4	14	28	56	74
s5	13	32	49	69

8.5.1 Data Entry Format

The data set has been saved under the name: **EX8b.SAV**.

Variables	Column(s)	Code
Sex	1	1 = men, 2 = women
T1	2	Running speed in sec
T2	3	Running speed in sec
T3	4	Running speed in sec
T4	5	Running speed in sec

8.5.2 Windows Method

1. To create the four variables of **FIFTY, ONEHUND, TWOHUND,** and **THREEHUN,** repeat step 1 to step 3 in Subsection 8.4.2.

2. From the menu bar, click **Analyze**, then **General Linear Model**, and then **Multivariate**. The following **Multivariate** window will open:

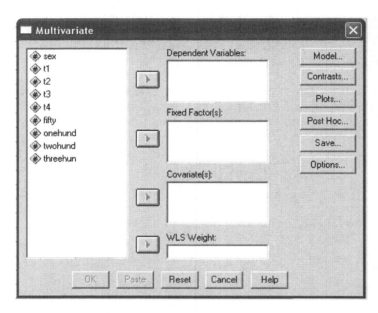

3. Transfer the four variables of **FIFTY, ONEHUND, TWOHUND,** and **THREEHUN** to the **Dependent Variables** field by clicking (high-

lighting) these four variables and then clicking 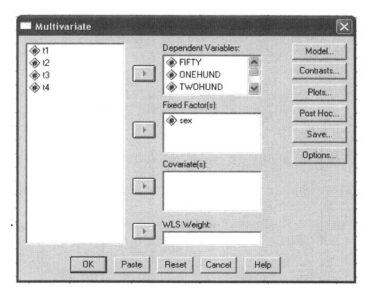. Transfer the **SEX** variable to the **Fixed Factor(s)** field by clicking (highlighting) this variable and clicking [▶].

4. To obtain the estimated marginal means for the four dependent variables for males and females, click Options... . When the **Multivariate: Options** window opens, click (highlight) **SEX** in the **Factor(s) and Factor Interactions** field and then click ▶ to transfer this factor to the **Display Means for** field. Click Continue to return to the **Multivariate** window.

5. When the **Multivariate** window opens, click OK to complete the analysis. See Table 8.2 for the results.

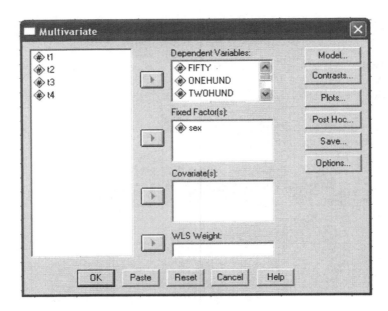

8.5.3 SPSS Syntax Method

COMPUTE FIFTY=(T1-6).
COMPUTE ONEHUND=(T2-12).
COMPUTE TWOHUND=(T3-25).
COMPUTE THREEHUN=(T4-40).

GLM FIFTY ONEHUND TWOHUND THREEHUN BY SEX
/EMMEANS=TABLES(SEX).

8.5.4 SPSS Output

TABLE 8.2

General Linear Model
Two-Sample GLM Output

Between-Subjects Factors			
		Value Label	N
Sex	1.00	Male	5
	2.00	Female	5

Multivariate Tests[b]

Effect		Value	F	Hypothesis df	Error df	Sig.
Intercept	Pillai's Trace	.997	358.769[a]	4.000	5.000	.000
	Wilks' Lambda	.003	358.769[a]	4.000	5.000	.000
	Hotelling's Trace	287.015	358.769[a]	4.000	5.000	.000
	Roy's Largest Root	287.015	358.769[a]	4.000	5.000	.000
Sex	Pillai's Trace	.971	41.742[a]	4.000	5.000	.000
	Wilks' Lambda	.029	41.742[a]	4.000	5.000	.000
	Hotelling's Trace	33.394	41.742[a]	4.000	5.000	.000
	Roy's Largest Root	33.394	41.742[a]	4.000	5.000	.000

[a]Exact statistic.
[b]Design: Intercept + sex.

Tests of between-Subjects Effects

Source	Dependent Variable	Type III Sum of Squares	df	Mean Square	F	Sig.
Corrected Model	FIFTY	67.600[a]	1	67.600	135.200	.000
	ONEHUND	250.000[b]	1	250.000	67.568	.000
	TWOHUND	36.100[c]	1	36.100	1.220	.302
	THREEHUN	250.000[d]	1	250.000	24.272	.001
Intercept	FIFTY	270.400	1	270.400	540.800	.000
	ONEHUND	1488.400	1	1488.400	402.270	.000
	TWOHUND	3572.100	1	3572.100	120.679	.000
	THREEHUN	8179.600	1	8179.600	794.136	.000
Sex	FIFTY	67.600	1	67.600	135.200	.000
	ONEHUND	250.000	1	250.000	67.568	.000
	TWOHUND	36.100	1	36.100	1.220	.302
	THREEHUN	250.000	1	250.000	24.272	.001
Error	FIFTY	4.000	8	.500		
	ONEHUND	29.600	8	3.700		
	TWOHUND	236.800	8	29.600		
	THREEHUN	82.400	8	10.300		
Total	FIFTY	342.000	10			
	ONEHUND	1768.000	10			
	TWOHUND	3845.000	10			
	THREEHUN	8512.000	10			
Corrected Total	FIFTY	71.600	9			
	ONEHUND	279.600	9			
	TWOHUND	272.900	9			
	THREEHUN	332.400				

[a]R Squared = .944 (Adjusted R Squared = .937).
[b]R Squared = .894 (Adjusted R Squared = .881).
[c]R Squared = .132 (Adjusted R Squared = .024).
[d]R Squared = .752 (Adjusted R Squared = .721).

Estimated Marginal Means

Sex

Dependent Variable	Sex	Mean	Standard Error	95% Confidence Interval	
				Lower Bound	Upper Bound
FIFTY	Male	2.600	.316	1.871	3.329
	Female	7.800	.316	7.071	8.529
ONEHUND	Male	7.200	.860	5.216	9.184
	Female	17.200	.860	15.216	19.184
TWOHUND	Male	17.000	2.433	11.389	22.611
	Female	20.800	2.433	15.189	26.411
THREEHUN	Male	23.600	1.435	20.290	26.910
	Female	33.600	1.435	30.290	36.910

8.5.5   Results and Interpretation

The **Estimated Marginal Means** table presents the mean running times for men and women, for each of the four running events after the normative values have been subtracted. Both samples exceed the norms for all four running events.

The **Multivariate Tests** table (Pillai's, Hotelling's, Wilks', and Roy's) tests the hypothesis that men and women do not differ significantly on overall running time. The significance level is based on the F distribution with 4 and 5 degrees of freedom. The observed significance levels for all four multivariate tests are small ($p < .001$), so the null hypothesis that men and women performed similarly on the running events is rejected (e.g., multivariate Pillai $F(4,5) = 41.74$, $p < .001$).

The **Tests of Between-Subjects Effects** table presents the univariate test of sex difference for the individual running events. The obtained F-values are equivalent to those obtained from the one-way analysis of variance. *In the case where the comparison is between two groups, the F-values are the square of the two-sample t-values.* The results indicate significant sex differences in running times for three of the four running events (FIFTY: male: M = 2.60, female: M = 7.80; ONEHUND: male: M = 7.20, female: M = 17.20; THREEHUN: male: M = 23.60, female: M = 33.60); there is no significant sex difference for the 200-yd dash.

8.6   Example 3 — GLM: 2 × 2 Factorial Design

In Example 2, the researcher used the GLM two-sample test to test for sex differences in running times across four running events. Suppose that in addition to sex differences, the researcher was also interested in whether the subjects' ethnicity (white vs. nonwhite) would make a difference to their running times. In particular, the researcher was interested in the interaction between subjects' sex and ethnicity in influencing their running times. The running times for the four groups (men-white, men-nonwhite, women-white, women-nonwhite) for the four events are presented in the following.

	Running Events			
	50 yd	100 yd	200 yd	300 yd
Men				
White				
s1	9	20	42	67
s2	9	19	45	66
s3	8	17	38	62
s4	9	18	46	64
s5	8	22	39	59
Nonwhite				
s1	6	17	39	64
s2	6	16	42	63
s3	5	14	35	59
s4	6	15	43	61
s5	5	19	36	56
Women				
White				
s1	15	30	41	77
s2	14	29	44	76
s3	13	27	39	72
s4	14	28	56	74
s5	13	32	49	69
Nonwhite				
s1	20	35	46	82
s2	19	34	49	81
s3	18	32	44	77
s4	19	32	61	79
s5	18	37	54	74

8.6.1 Data Entry Format

The data set has been saved under the name: **EX8c.SAV**.

Variables	Column(s)	Code
Sex	1	1 = men, 2 = women
Ethnic	2	1 = white, 2 = nonwhite
T1	3	Running speed in sec
T2	4	Running speed in sec
T3	5	Running speed in sec
T4	6	Running speed in sec

8.6.2 Windows Method

1. To create the four variables of **FIFTY, ONEHUND, TWOHUND,** and **THREEHUN** repeat step 1 to step 3 in Subsection 8.4.2.

2. From the menu bar, click **Analyze**, then **General Linear Model**, and then **Multivariate**. The following **Multivariate** window will open:

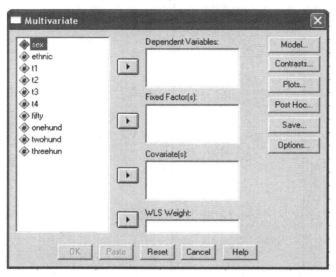

3. Transfer the four variables of **FIFTY, ONEHUND, TWOHUND,** and **THREEHUN** to the **Dependent Variables** field by clicking (high-lighting) these four variables and then clicking [▶]. Transfer the **SEX** and **ETHNIC** variables to the **Fixed Factor(s)** field by clicking (highlighting) these variables and then clicking [▶].

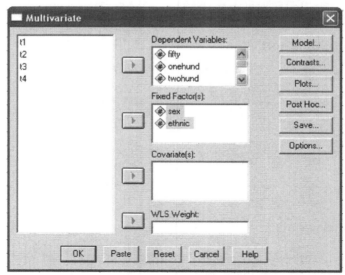

4. To obtain the estimated marginal means for the four dependent variables for the four experimental groups (men-white, men-non-white, women-white, women-nonwhite), click Options... . When the **Multivariate: Options** window opens, click (highlight) **SEX, ETHNIC, SEX*ETHNIC** in the **Factor(s) and Factor Interactions** field, and then click ► to transfer these factors and factor interaction to the **Display Means for** field. Click Continue to return to the **Multivariate** window.

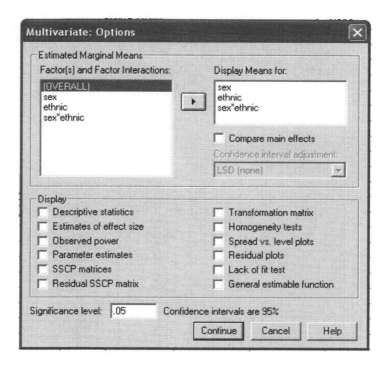

5. When the **Multivariate** window opens, click OK to complete the analysis. See Table 8.3 for the results.

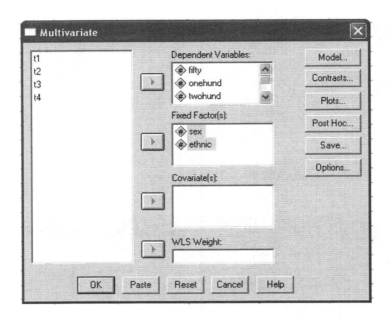

8.6.3 SPSS Syntax Method

COMPUTE FIFTY=(T1-6).
COMPUTE ONEHUND=(T2-12).
COMPUTE TWOHUND=(T3-25).
COMPUTE THREEHUN=(T4-40).

GLM FIFTY ONEHUND TWOHUND THREEHUN BY SEX ETHNIC
/EMMEANS=TABLES(SEX)
/EMMEANS=TABLES(ETHNIC)
/EMMEANS=TABLES(SEX*ETHNIC).

8.6.4 SPSS Output

TABLE 8.3

General Linear Model
Two × Two GLM Output

Between -Subjects Factors			
		Value Label	N
Sex	1.00	male	10
	2.00	female	10
Ethnic	1.00	WHITE	10
	2.00	NONWHITE	10

Multivariate Tests[b]						
Effect		Value	F	Hypothesis df	Error df	Sig.
Intercept	Pillai's Trace	.996	890.965[a]	4.000	13.000	.000
	Wilks' Lambda	.004	890.965[a]	4.000	13.000	.000
	Hotelling's Trace	274.143	890.965[a]	4.000	13.000	.000
	Roy's Largest Root	274.143	890.965[a]	4.000	13.000	.000
Sex	Pillai's Trace	.992	392.434[a]	4.000	13.000	.000
	Wilks' Lambda	.008	392.434[a]	4.000	13.000	.000
	Hotelling's Trace	120.749	392.434[a]	4.000	13.000	.000
	Roy's Largest Root	120.749	392.434[a]	4.000	13.000	.000
Ethnic	Pillai's Trace	.658	6.254[a]	4.000	13.000	.005
	Wilks' Lambda	.342	6.254[a]	4.000	13.000	.005
	Hotelling's Trace	1.924	6.254[a]	4.000	13.000	.005
	Roy's Largest Root	1.924	6.254[a]	4.000	13.000	.005
Sex * ethnic	Pillai's Trace	.968	99.122[a]	4.000	13.000	.000
	Wilks' Lambda	.032	99.122[a]	4.000	13.000	.000
	Hotelling's Trace	30.499	99.122[a]	4.000	13.000	.000
	Roy's Largest Root	30.499	99.122[a]	4.000	13.000	.000

[a]Exact statistic.
[b]Design: Intercept + sex + ethnic + sex * ethnic.

Tests of between-Subjects Effects

Source	Dependent Variable	Type III Sum of Squares	df	Mean Square	F	Sig.
Corrected	FIFTY	508.200[a]	3	169.400	338.800	.000
Model	ONEHUND	1046.150[b]	3	348.717	89.415	.000
	TWOHUND	389.200[c]	3	129.733	4.383	.020
	THREEHUN	1065.000[d]	3	355.000	34.466	.000
Intercept	FIFTY	649.800	1	649.800	1299.600	.000
	ONEHUND	3200.450	1	3200.450	820.628	.000
	TWOHUND	7527.200	1	7527.200	254.297	.000
	THREEHUN	16936.200	1	16936.200	1644.291	.000
Sex	FIFTY	423.200	1	423.200	846.400	.000
	ONEHUND	966.050	1	966.050	247.705	.000
	TWOHUND	304.200	1	304.200	10.277	.006
	THREEHUN	980.000	1	980.000	95.146	.000
Ethnic	FIFTY	5.000	1	5.000	10.000	.006
	ONEHUND	4,050	1	4.050	1.038	.323
	TWOHUND	5.000	1	5.000	.169	.687
	THREEHUN	5.000	1	5.000	.485	.496
Sex*ethnic	FIFTY	80.000	1	80.000	160.000	.000
	ONEHUND	76.050	1	76.050	19.500	.000
	TWOHUND	80.000	1	80.000	2.703	.120
	THREEHUN	80.000	1	80.000	7.767	.013
Error	FIFTY	8.000	16	.500		
	ONEHUND	62.400	16	3.900		
	TWOHUND	473.600	16	29.600		
	THREEHUN	164.800	16	10.300		
Total	FIFTY	1166.000	20			
	ONEHUND	4309.000	20			
	TWOHUND	8390.000	20			
	THREEHUN	18166.000	20			
Corrected	FIFTY	516.200	19			
Total	ONEHUND	1108.550	19			
	TWOHUND	862.800	19			
	THREEHUN	1229.800	19			

[a]R Squared = .985 (Adjusted R Squared = .982).
[b]R Squared = .944 (Adjusted R Squared = .933).
[c]R Squared = .451 (Adjusted R Squared = .348).
[d]R Squared = .866 (Adjusted R Squared = .841).

Estimated Marginal Means

1. Sex

Dependent Variable	Sex	Mean	Standard Error	95% Confidence Interval	
				Lower Bound	Upper Bound
FIFTY	Male	1.100	.224	.626	1.574
	Female	10.300	.224	9.826	10.774
ONEHUND	Male	5.700	.624	4.376	7.024
	Female	19.600	.624	18.276	20.924
TWOHUND	Male	15.500	1.720	11.853	19.147
	Female	23.300	1.720	19.653	26.947
THREEHUN	male	22.100	1.015	19.949	24.251
	female	36.100	1.015	33.949	38.251

2. Ethnic

Dependent Variable	Ethnic	Mean	Standard Error	95% Confidence Interval	
				Lower Bound	Upper Bound
FIFTY	WHITE	5.200	.224	4.726	5.674
	NONWHITE	6.200	.224	5.726	6.674
ONEHUND	WHITE	12.200	.624	10.876	13.524
	NONWHITE	13.100	.624	11.776	14.424
TWOHUND	WHITE	18.900	1.720	15.253	22.547
	NONWHITE	19.900	1.720	16.253	23.547
THREEHUN	WHITE	28.600	1.015	26.449	30.751
	NONWHITE	29.600	1.015	27.449	31.751

3. Sex*Ethnic

Dependent Variable	Sex	Ethnic	Mean	Standard Error	95% Confidence Interval	
					Lower Bound	Upper Bound
FIFTY	Male	WHITE	2.600	.316	1.930	3.270
		NONWHITE	−.400	.316	−1.070	.270
	Female	WHITE	7.800	.316	7.130	8.470
		NONWHITE	12.800	.316	12.130	13.470
ONEHUND	Male	WHITE	7.200	.883	5.328	9.072
		NONWHITE	4.200	.883	2.328	6.072
	Female	WHITE	17.200	.883	15.328	19.072
		NONWHITE	22.000	.883	20.128	23.872
TWOHUND	Male	WHITE	17.000	2.433	11.842	22.158
		NONWHITE	14.000	2.433	8.842	19.158
	Female	WHITE	20.800	2.433	15.642	25.958
		NONWHITE	25.800	2.433	20.642	30.958
THREEHUN	Male	WHITE	23.600	1.435	20.557	26.643
		NONWHITE	20.600	1.435	17.557	23.643
	Female	WHITE	33.600	1.435	30.557	36.643
		NONWHITE	38.600	1.435	35.557	41.643

8.6.5   Results and Interpretation

The **Estimated Marginal Means** table presents the means of the overall running times for the four running events as a function of subjects' **SEX**, **ETHNICITY**, and **SEX*ETHNICITY** interaction.

The **Multivariate Tests** table presents the multivariate tests of significance for the main effects of the between-groups variables of **SEX** and **ETHNICITY**, and the **SEX*ETHNICITY** interaction. For all three effects, the observed significance levels for the four multivariate tests (Pillai's, Wilks', Hotelling's, and Roy's) are small. Thus, their associated null hypotheses (no sex difference, no ethnicity difference, and no sex*ethnicity interaction) are rejected. The results from the multivariate tests have been converted to approximate F-values and can be interpreted in the same way that F-values from one-way ANOVA are interpreted. The multivariate tests for both **SEX** (multivariate Pillai F(4,13) = 392.43, $p < .001$) and **ETHNICITY** (multivariate Pillai F(4,13) = 6.25, $p < .01$) are statistically significant, indicating that men (M = 11.1), when compared to women (M = 22.32), and whites (M = 13.18), when compared to nonwhites (M = 17.2), differed significantly in their overall running times.

The **Tests of Between-Subjects Effects** can be examined for significant **SEX** and **ETHNICITY** differences for each of the four running events. For **SEX**, the results show significant sex differences for all four running events ($p < .01$) (**FIFTY**: men: M = 1.10, women: M = 10.30; **ONEHUND**: men: M = 5.70, women: M = 19.60; **TWOHUND**: men: M = 15.50, women: M = 23.30; **THREEHUN**: men: M = 22.10, women: M = 36.10). For **ETHNICITY**, the results show significant ethnicity difference for only the FIFTY-yd dash (white: M = 5.20, nonwhite: M = 6.20; F(1,16) = 10.00, $p < .01$).

Given that the **SEX*ETHNICITY** interaction is significant, the **Tests of Between-Subjects Effects** table can be examined to see which of the four running events are significantly affected by this interaction. The univariate F-tests for the **SEX*ETHNICITY** interaction are identical to the F ratios generated from the **SEX*ETHNICITY** interaction in a two-way ANOVA. The univariate results show significant **SEX*ETHNICITY** interaction effect for three of the four running events ($p < .01$). Thus, the subjects' performances on the 50-yd dash, the 100-yd dash, and the 300-yd dash are dependent on the joint effects of their sex and ethnicity.

To interpret the **SEX*ETHNICITY** interactions, it would be useful to graph the means of the four running events (**FIFTY, ONEHUND, TWOHUND,** and **THREEHUN**) from the estimated marginal means table. The following **Windows Method** and the **Syntax File Method** will generate the graphs presented in Figure 8.1, Figure 8.2, Figure 8.3, and Figure 8.4.

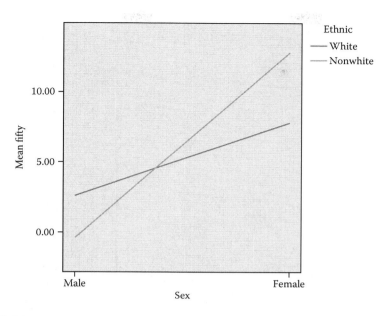

FIGURE 8.1

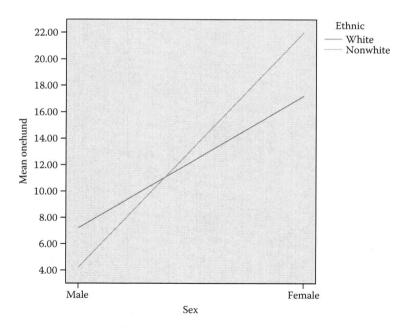

FIGURE 8.2

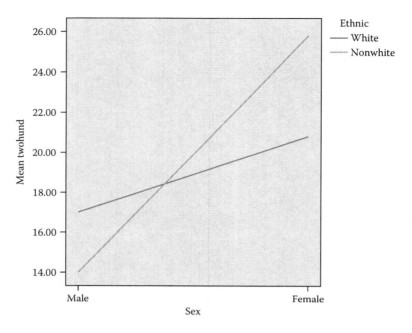

FIGURE 8.3

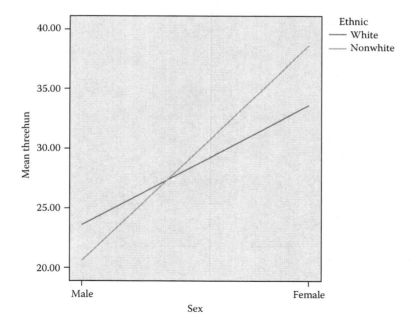

FIGURE 8.4

8.6.6 Windows Method (Profile Plot)

1. Open the **Multivariate** window and click **Plots...** . The following **Multivariate: Profile Plots** window will open. Transfer the **SEX** variable to the **Horizontal Axis** field by clicking (highlighting) the variable and then clicking **▶** . Transfer the **ETHNIC** variable to the **Separate Lines** field by clicking (highlighting) the variable and then clicking **▶** . Next, click **Add** to transfer the **SEX*ETHNIC** interaction to the **Plots** field. When this is done, click **Continue** .

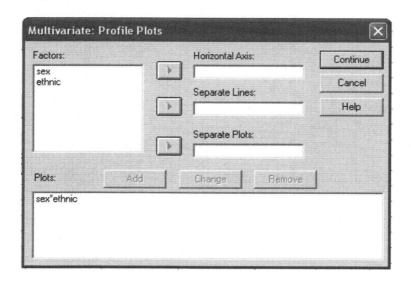

2. When the **Multivariate** window opens, click **OK** to plot the graphs.

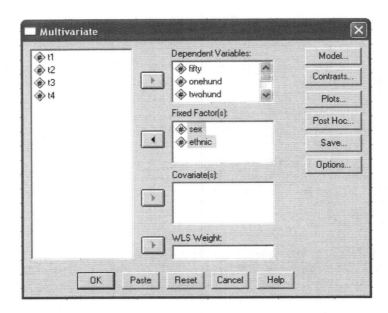

8.6.7   SPSS Syntax Method (Profile Plot)

GRAPH
/LINE(MULTIPLE)MEAN(FIFTY) BY SEX BY ETHNIC.
GRAPH
/LINE(MULTIPLE)MEAN(ONEHUND) BY SEX BY ETHNIC.
GRAPH
/LINE(MULTIPLE)MEAN(TWOHUND) BY SEX BY ETHNIC.
GRAPH
/LINE(MULTIPLE)MEAN(THREEHUN) BY SEX BY ETHNIC.

8.6.8   Results and Interpretation

Using the 100-yd dash as an example, the **SEX*ETHNICITY** interaction can be interpreted as follows. The effect of subjects' sex on the running times in the 100-yd dash is dependent on the ethnicity of the subjects, such that for men, nonwhites ran faster than whites; for women, the effect is opposite with whites running faster than nonwhites.

Post hoc comparisons analysis can also be used to identify which of the four groups (male-white, male-nonwhite, female-white, female-nonwhite) generated from the **SEX*ETHNICITY** interaction is significantly different from each other in the 100-yd dash. The following **Windows Method** and the **Syntax File Method** will accomplish this analysis.

8.6.9 Data Transformation — Windows Method and SPSS Syntax Method

8.6.9.1 Data Transformation

1. The first step is to create a new grouping variable called **GROUP** that contains the four levels (white male, nonwhite male, white female, and nonwhite female) generated by the **SEX*ETHNIC** interaction. From the menu bar, click **Transform** and then **Compute**. The following **Compute Variable** window will open:

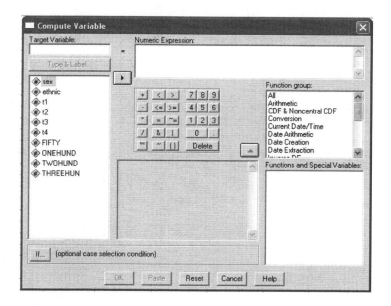

2. Click **If...** to open the **Compute Variable: If Cases** window. Ensure that **Include if case satisfies condition** is checked. To create the first group (**male-white**), type **sex=1 and ethnic=1** in the field. Click **Continue**.

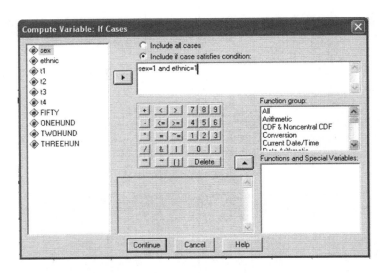

3. Because the **male-white** group is the first of four levels within a new grouping variable called **GROUP**, this level will be coded **1** within the **GROUP** variable. When the following window opens, type **GROUP** in the **Target Variable** field, and **1** in the **Numeric Expression** field. Click [OK] to create the first level of **male-white** (coded **1**) within the new grouping variable of **GROUP**.

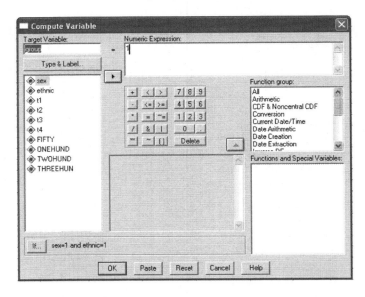

4. Repeat steps 1 to 3 to create the other three levels: **male-nonwhite** (coded **2**), **female-white** (coded **3**), and **female-nonwhite** (coded **4**).

5. To aid interpretation of the obtained results, **Value Labels** in the data set should be activated and labels attached to the numerical codes for the four levels. To do this, open the data set, and under **Variable View**, click the **Values** field for the **GROUP** variable. Type in the value labels as shown in the **Value Labels** window in the following:

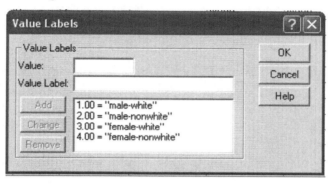

8.6.9.2 Post Hoc Comparisons

Once the four levels (**male-white, male-nonwhite, female-white, female-nonwhite**) have been created, **Scheffé** post hoc comparisons can be conducted to test for differences (simple effects) between these four levels.

8.6.9.2.1 Post Hoc Comparisons (Windows Method)

From the menu bar, click **Analyze**, then **Compare Means**, and then **One-Way ANOVA**. The following **One-Way ANOVA** window will open. Note that the listing of variables now includes the newly created variable of **GROUP** which contains the four levels of male-white, male-nonwhite, female-white, and female-nonwhite.

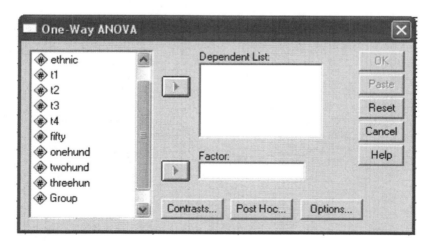

1. Transfer the dependent variable of **ONEHUND** to the **Dependent List** field by clicking (highlighting) the variable and then clicking

   ▶ . Transfer the independent variable of **GROUP** to the **Factor**

   field by clicking (highlighting) the variable and then clicking ▶ .
   To do post hoc comparisons for the running times in the 100-yd dash

   between the four levels, click Post Hoc... .

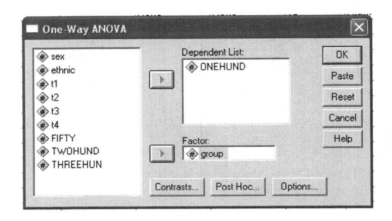

2. When the following **One-Way ANOVA: Post Hoc Multiple Comparisons** window opens, check the **Scheff**é box to run the Scheffé

   post hoc test. Next, click Continue .

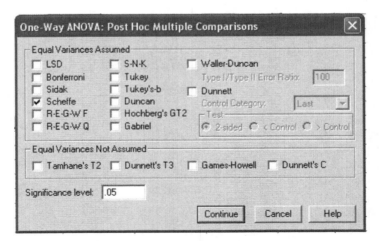

3. When the **One-Way ANOVA** window opens, run the analysis by clicking [OK] . See Table 8.4 for the results.

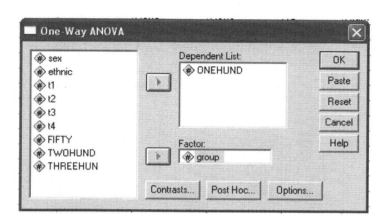

8.6.9.2.2 Post Hoc Comparisons (SPSS Syntax Method)

IF (SEX EQ 1 AND ETHNIC EQ 1) GROUP=1.
IF (SEX EQ 1 AND ETHNIC EQ 2) GROUP=2.
IF (SEX EQ 2 AND ETHNIC EQ 1) GROUP=3.
IF (SEX EQ 2 AND ETHNIC EQ 2) GROUP=4.
VALUE LABELS GROUP 1 'MALE-WHITE' 2 'MALE-NONWHITE'
3 'FEMALE-WHITE' 4 'FEMALE-NONWHITE'.

ONEWAY ONEHUND BY GROUP (1,4)
/RANGES=SCHEFFE(.05).

8.6.10   SPSS Output

TABLE 8.4

Scheffé Post Hoc Comparisons Output

Multiple Comparisons

Dependent Variable: ONEHUND

Scheffe

(I) GROUP	(J) GROUP	Mean Difference (I – J)	Standard Error	Sig.	95% Confidence Interval Lower Bound	Upper Bound
MALE-WHITE	MALE-NONWHITE	3.0000	1.2490	.167	−.8933	6.8933
	FEMALE-WHITE	−10.0000*	1.2490	.000	−13.8933	−6.1067
	FEMALE-NONWHITE	−14.8000*	1.2490	.000	−18.6933	−10.9067
MALE-NONWHITE	MALE-WHITE	−3.0000	1.2490	.167	−6.8933	.8933
	FEMALE-WHITE	−13.0000*	1.2490	.000	−16.8933	−9.1067
	FEMALE-NONWHITE	−17.8000*	1.2490	.000	−21.6933	−13.9067
FEMALE-WHITE	MALE-WHITE	10.0000*	1.2490	.000	6.1067	13.8933
	MALE-NONWHITE	13.0000*	1.2490	.000	9.1067	16.8933
	FEMALE-NONWHITE	−4.8000*	1.2490	.013	−8.6933	−.9067
FEMALE-NONWHITE	MALE-WHITE	14.8000*	1.2490	.000	10.9067	18.6933
	MALE-NONWHITE	17.8000*	1.2490	.000	13.9067	21.6933
	FEMALE-WHITE	4.8000*	1.2490	.013	.9067	8.6933

*The mean difference is significant at the .05 level.

8.6.11   Results and Interpretation

Results from the Scheffé comparisons indicate that white females (M = 17.20) and nonwhite females (M = 22.00) ran significantly slower than white males (M = 7.20) and nonwhite males (M = 4.20). Whereas there was no significant difference in running times between white males and nonwhite males, the results show that nonwhite females ran significantly slower than white females.

9

General Linear Model: Repeated Measures Analysis

9.1 Aim

When a subject is tested on the same variable over time, it is a repeated measures design. Although the advantages of repeated measurements are obvious (e.g., they require fewer subjects per experiment, and they eliminate between-subjects differences from the experimental error), they violate the most important assumption of multivariate analysis — *independence*. GLM for repeated measures is a special procedure that can account for this dependence and still test for differences across individuals for the set of dependent variables.

9.2 Example 1 — GLM: One-Way Repeated Measures

An investigator is interested in studying how exposure to different levels of temperature influences problem-solving ability. Five subjects were asked to solve a series of mathematical problems under four different temperature conditions. The temperature in the room in which the subjects were tested was originally set at 35°C, with the temperature dropping by 5°C every 5 min to a minimum of 20°C. Thus, each subject was required to solve a set of mathematical problems under the four temperature conditions of 35°C, 30°C, 25°C, and 20°C. The investigator expected that the greatest number of errors will be made when the room temperature is 35°C, fewer errors will be made when the room temperature is 30°C, still fewer when the room temperature is 25°C, and the fewest errors will be made when the room temperature is 20°C. The data given in the following represent the number of errors made by five subjects across the four conditions.

| | Room Temperature | | | |
Subjects	35°C	30°C	25°C	20°C
s1	7	5	2	3
s2	8	7	5	4
s3	6	5	3	3
s4	8	8	4	2
s5	5	4	3	2

9.2.1   Data Entry Format

The data set has been saved under the name **EX9a.SAV**.

Variables	Column(s)	Code
Temp1	1	Number of errors
Temp2	2	Number of errors
Temp3	3	Number of errors
Temp4	4	Number of errors

9.2.2   Windows Method

1. From the menu bar, click **Analyze**, then **General Linear Model**, and then **Repeated Measures**. The following **Repeated Measures Define Factor(s)** window will open.

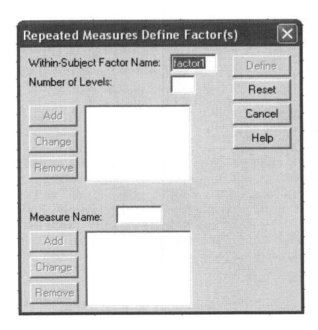

2. In the **Within-Subject Factor Name** field, type the name **TEMP** to denote the name of the within-subject factor. In the **Number of Levels** field, type **4** to denote the four levels of the within-subject factor **TEMP** (Temp1, Temp2, Temp3, and Temp4). Click ▭ Add ▭ to transfer the **TEMP** factor to the **ADD** field. In the **Measure Name** field, type the name of the dependent measure: **ERRORS**, and then click ▭ Add ▭ .

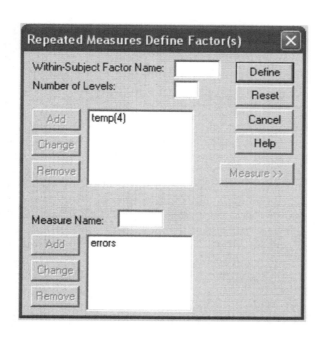

3. Click ▭ Define ▭ to open the **Repeated Measures** window. Transfer the four variables of Temp1, Temp2, Temp3, and Temp4 to the **Within-Subjects Variables (temp)** field by clicking these variables (highlight) and then clicking ▭ ▶ ▭ .

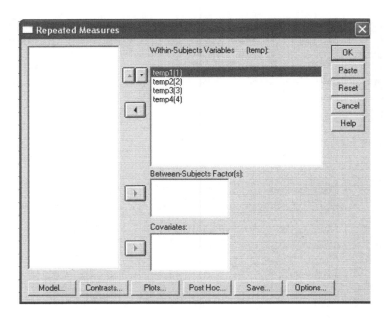

4. To test for differences among the four levels (Temp1, Temp2, Temp3, and Temp4) of the within-subjects factor **TEMP**, click Contrasts... to open the **Repeated Measures: Contrasts** window. Select a contrast from the **Contrast** drop-down list and right-click the mouse button on the contrast to see its description. For this example, choose **Repeated** as the contrast. This contrast compares the mean of each level (except the last) to the mean of the subsequent level. Click Continue to return to the **Repeated Measures** window.

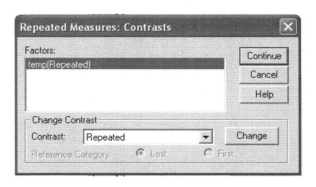

5. In the **Repeated Measures** window, click Options... to open the **Repeated Measures: Options** window. To display means for the four levels of **TEMP** (Temp1, Temp2, Temp3, and Temp4), click (highlight) **TEMP** in the **Factor(s) and Factor Interactions** field, and then

click ▶ to transfer this variable to the **Display Means for** field. To compute **pairwise comparisons** of the means for Temp1, Temp2, Temp3, and Temp4, check the **Compare main effects** cell. However, because of inflated Type I error rate due to the multiple comparisons, a **Bonferroni** type adjustment should be made. From the **Confidence interval adjustment** field, select **Bonferroni** from the drop-down

list. Click Continue to return to the **Repeated Measures** window.

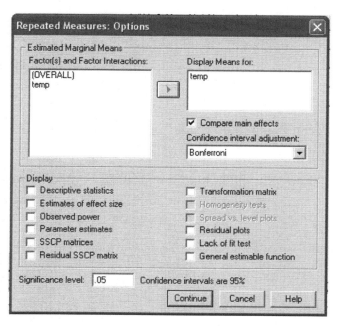

6. When the **Repeated Measures** window opens, click OK to complete the analysis. See Table 9.1 for the results.

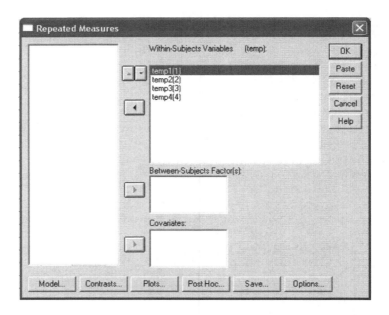

9.2.3   SPSS Syntax Method

GLM TEMP1 TO TEMP4
/WSFACTOR=TEMP 4 REPEATED
/MEASURE=ERRORS
/EMMEANS=TABLES(TEMP) COMPARE ADJ(BONFERRONI).

Note: As the investigator expected fewer errors to be made when the room temperature dropped from 35 to 20°C, specific comparisons or contrasts between the temperature conditions can be made. In this example, the **Repeated** contrast has been used to compare the mean of each level (except the last) to the mean of the subsequent level. SPSS offers the following contrast types for GLM Repeated Measures analysis:

- **Deviation.** Compares the mean of each level (except a reference category) to the mean of all of the levels (grand mean). The levels of the factor can be in any order.
- **Simple.** Compares the mean of each level to the mean of a specified level. This type of contrast is useful when there is a control group. You can choose the first or last category as the reference.
- **Difference.** Compares the mean of each level (except the first) to the mean of previous levels.
- **Helmert.** Compares the mean of each level of the factor (except the last) to the mean of subsequent levels.

- **Repeated.** Compares the mean of each level (except the last) to the mean of the subsequent level.
- **Polynomial.** Compares the linear effect, quadratic effect, cubic effect, and so on. The first degree of freedom contains the linear effect across all categories; the second degree of freedom, the quadratic effect; and so on. These contrasts are often used to estimate polynomial trends.

9.2.4 SPSS Output

TABLE 9.1

GLM One-Way Repeated Measures Output

Within-Subjects Factors

Measure: MEASURE_1

TEMP	Dependent Variable
1	TEMP1
2	TEMP2
3	TEMP3
4	TEMP4

Multivariate Tests[b]

	Effect	Value	F	Hypothesis df	Error df	Sig.
TEMP	Pillai's Trace	.985	43.500[a]	3.000	2.000	.023
	Wilks' Lambda	.015	43.500[a]	3.000	2.000	.023
	Hotelling's Trace	65.250	43.500[a]	3.000	2.000	.023
	Roy's Largest Root	65.250	43.500[a]	3.000	2.000	.023

[a] Exact statistic.
[b] Design: Intercept, Within Subjects Design: TEMP.

Mauchly's Test of Sphericity[b]

Measure: MEASURE_1

Within Subjects Effect	Mauchly's W	Approx. Chi-Square	df	Sig.	Epsilon[a]		
					Greenhouse-Geisser	Huynh-Feldt	Lower-bound
TEMP	.054	7.965	5	.182	.571	.958	.333

Note: Tests the null hypothesis that the error covariance matrix of the orthonormalized transformed dependent variables is proportional to an identity matrix.

[a] May be used to adjust the degrees of freedom for the averaged tests of significance. Corrected tests are displayed in the Tests of Within-Subjects Effects table.
[b] Design: Intercept, Within Subjects Design: TEMP.

Tests of within-Subjects Effects

Measure: MEASURE_1

Source		Type III Sum of Squares	df	Mean Square	F	Sig.
TEMP	Sphericity Assumed	54.600	3	18.200	24.539	.000
	Greenhouse–Geisser	54.600	1.714	31.853	24.539	.001
	Huynh–Feldt	54.600	2.874	18.995	24.539	.000
	Lower-bound	54.600	1.000	54.600	24.539	.008
Error(TEMP)	Sphericity Assumed	8.900	12	.742		
	Greenhouse–Geisser	8.900	6.857	1.298		
	Huynh–Feldt	8.900	11.498	.774		
	Lower-bound	8.900	4.000	2.225		

Tests of within-Subjects Contrasts

Measure: MEASURE_1

Source	TEMP	Type III Sum of Squares	df	Mean Square	F	Sig.
TEMP	Level 1 vs. Level 2	5.000	1	5.000	10.000	.034
	Level 2 vs. Level 3	28.800	1	28.800	22.154	.009
	Level 3 vs. Level 4	1.800	1	1.800	1.385	.305
Error(TEMP)	Level 1 vs. Level 2	2.000	4	.500		
	Level 2 vs. Level 3	5.200	4	1.300		
	Level 3 vs. Level 4	5.200	4	1.300		

Tests of between-Subjects Effects

Measure: MEASURE_1

Transformed Variable: Average

Source	Type III Sum of Squares	df	Mean Square	F	Sig.
Intercept	110.450	1	110.450	105.820	.001
Error	4.175	4	1.044		

Estimated Marginal Means

2. TEMP

Estimates

Measure: ERRORS

TEMP	Mean	Standard Error	95% Confidence Interval	
			Lower Bound	Upper Bound
1	6.800	.583	5.181	8.419
2	5.800	.735	3.760	7.840
3	3.400	.510	1.984	4.816
4	2.800	.374	1.761	3.839

Pairwise Comparisons

Measure: ERRORS

(I) TEMP	(J) TEMP	Mean Difference (I − J)	Standard Error	Sig.[a]	95% Confidence Interval for Difference[a]	
					Lower Bound	Upper Bound
1	2	1.000	.316	.205	−.534	2.534
	3	3.400*	.510	.016	.926	5.874
	4	4.000*	.548	.011	1.343	6.657
2	1	−1.000	.316	.205	−2.534	.534
	3	2.400	.510	.056	−.074	4.874
	4	3.000	.775	.108	−.758	6.758
3	1	−3.400*	.510	.016	−5.874	−.926
	2	−2.400	.510	.056	−4.874	.074
	4	.600	.510	1.000	−1.874	3.074
4	1	−4.000*	.548	.011	−6.657	−1.343
	2	−3.000	.775	.108	−6.758	.758
	3	−.600	.510	1.000	−3.074	1.874

Note: Based on estimated marginal means.

*The mean difference is significant at the .05 level.

[a]Adjustment for multiple comparisons: Bonferroni.

Multivariate Tests

	Value	F	Hypothesis df	Error df	Sig.
Pillai's trace	.985	43.500[a]	3.000	2.000	.023
Wilks' lambda	.015	43.500[a]	3.000	2.000	.023
Hotelling's trace	65.250	43.500[a]	3.000	2.000	.023
Roy's largest root	65.250	43.500[a]	3.000	2.000	.023

Note: Each F tests the multivariate effect of TEMP. These tests are based on the linearly independent pairwise comparisons among the estimated marginal means.

[a]Exact statistic.

9.2.5   Choosing Tests of Significance

The experiment involves testing the effect of the within-subjects variable of
TEMP on the number of errors made. In interpreting the results of the
analysis, the SPSS output allows the researcher to interpret either the **Multivariate Tests** results or the **Tests of Within-Subjects Effects** results. Choosing between the two depends on the outcome of the **Mauchly's Test of
Sphericity**, which tests the hypothesis that the covariance matrix has a constant variance on the diagonal and zeros off the diagonal, i.e., the correlations
between the variables are all roughly the same.

- If the **Mauchly's Test of Sphericity** is not significant (i.e., the
  assumption about the characteristics of the variance-covariance
  matrix is not violated), the **Tests of Within-Subjects Effects** can be
  used.
- If the **Mauchly's Test of Sphericity** is significant, the **Multivariate
  Tests** should be used.

However, if the researcher chooses to interpret the **Tests of Within-Subjects Effects** averaged F tests, then an adjustment to the numerator and
denominator degrees of freedom must be made. Both the numerator and
denominator degrees of freedom must be multiplied by the **Huynh-Feldt
epsilon** (presented as part of the **Tests of Within-Subjects Effects** table),
and the significance of the F ratio evaluated with the new degrees of freedom.

9.2.6   Results and Interpretation

The results from the analysis indicate that the **Mauchly's Sphericity Test** is
not significant (p = .158). Therefore, the **Tests of Within-Subjects Effects** can
be interpreted. The result indicates that the within-subjects variable of TEMP
is highly significant, $F(3,12)$ = 24.54, p < .001. That is, the number of errors
made by the subjects differed significantly as a function of the four temperature conditions. This is supported by the decrease in the mean number of
errors made as the room temperature decreased from 35 to 20°C.

Had the results of the **Mauchly's Sphericity Test** been significant, and had
the researcher decided to interpret the **Tests of Within-Subjects Effects**, then
an adjustment to the numerator and denominator degrees of freedom must
be made. This is achieved by multiplying these two values by the Huynh-
Feldt epsilon. Therefore, the adjusted numerator degrees of freedom is 3 ×
0.95816 = 2.87; the adjusted denominator degrees of freedom is 12 × 0.95816
= 11.50. The F ratio of 24.54 must then be evaluated with these new degrees
of freedom. The significance of each test when sphericity is assumed, or
when the Huynh–Feldt epsilon is used, is displayed under the **Tests of
Within-Subjects Effects** table.

As the within-subjects variable of **TEMP** is statistically significant, results from the **Repeated** contrast can be interpreted to determine which variables contributed to the overall difference. These results are presented under the heading of **Tests of Within-Subjects Contrasts**. The first contrast is between Level 1 (Temp1) vs. Level 2 (Temp2), which is statistically significant, $F(1,4) = 10.00$, $p < .05$. This means that the mean number of errors made in the 35°C condition (M = 6.80) is significantly greater than the mean number of errors made in the 30°C condition (M = 5.80). The second contrast is between Level 2 (Temp2) vs. Level 3 (Temp3), which is statistically significant, $F(1,4) = 22.15$, $p < .01$. This means that the mean number of errors made in the 30°C condition (M = 5.80) is significantly greater than the mean number of errors made in the 25°C condition (M = 3.40). The third contrast is between Level 3 (Temp3) vs. Level 4 (Temp4) and is insignificant, $F(1,4) = 1.39$, $p > .05$. This means that the difference in the mean number of errors made in the 25°C (M = 3.40) and 20°C (M = 2.80) conditions is due to chance variation.

The **Pairwise Comparisons** table presents all pairwise comparisons (with Bonferroni adjustment) between the four levels. The results indicate that the number of errors generated under Temp1 (35°C) is significantly greater than the number of errors generated in both Temp3 (25°C) and Temp4 (20°C). There is no significant difference in the number of errors generated under Temp1 (35°C) and Temp2 (30°C).

The **Multivariate Tests** statistics (Pillai's trace, Wilks' Lambda, Hotelling's trace, Roy's largest root) indicate that the overall difference in the number of errors generated across the four levels of temperature is statistically significant ($p < .05$).

9.3 Example 2 — GLM: Two-Way Repeated Measures (Doubly Multivariate Repeated Measures)

In the previous example, subjects were tested on one variable across four trials. In the following example, subjects will be tested on two variables, i.e., there will be two within-subjects variables.

An experimenter wants to study the effect of pain on memorization. Specifically, he wants to study the relationship between strength of electrical shock (low vs. high shocks) and type of items (easy, moderate, and difficult) on retention. The scores recorded for each subject are the subject's total number of correct responses on the easy, moderate, and difficult items when he/she is tested under low and high shocks.

Subject	Low Shock			High Shock		
	Easy	Moderate	Difficult	Easy	Moderate	Difficult
s1	83	88	100	75	60	40
s2	100	92	83	71	54	41
s3	100	97	100	50	47	43
s4	89	87	92	74	58	47
s5	100	96	90	83	72	62
s6	100	89	100	100	86	70
s7	92	94	92	72	63	44
s8	100	95	89	83	74	62
s9	50	60	54	33	25	38
s10	100	97	100	72	64	50

9.3.1   Data Entry Format

The data set has been saved under the name: EX9b.SAV

Variables	Column(s)	Code
LS_E	1	Number of correct responses
LS_M	2	Number of correct responses
LS_D	3	Number of correct responses
HS_E	4	Number of correct responses
HS_M	5	Number of correct responses
HS_D	6	Number of correct responses

9.3.2   Windows Method

1. From the menu bar, click **Analyze**, then **General Linear Model**, and then **Repeated Measures**. The following **Repeated Measures Define Factor(s)** window will open.

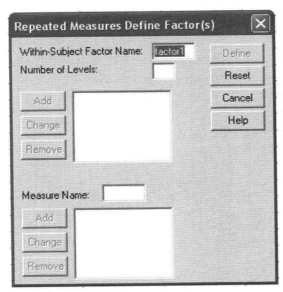

2. In the **Within-Subject Factor Name** field, type **SHOCK** to denote the name of the first within-subject factor. In the **Number of Levels** field, type **2** to denote the two levels of the **SHOCK** factor (Low Shock and High Shock). Click to transfer the **SHOCK** factor to the **ADD** field. Repeat this procedure to create the second within-subject factor of **ITEMS**. For this factor, there are three levels (Easy, Moderate, and Difficult). In the **Measure Name** field, type the name of the dependent measure: **CORRECT**, and then click

Add .

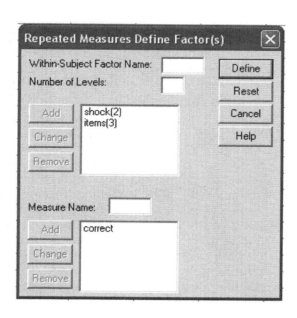

3. Click Define to open the **Repeated Measures** window. Transfer the six variables of LS_E, LS_M, LS_D, HS_E, HS_M, and _D (generated from the 2 (**SHOCK**) × 3 (**ITEMS**) factorial combination) to the **Within-Subjects Variables (shock,items)** cell by clicking these variables (highlighting) and then clicking [▶] .

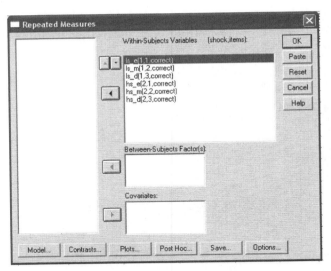

4. To test for differences among the six levels of LS_E, LS_M, L_D, HS_E, HS_M, and HS_D (generated from the factorial combination of the two within-subject factors of **SHOCK** and **ITEMS**), click

   [Contrasts...] to open the **Repeated Measures: Contrasts** window. Select **Repeated** as the contrast from the **Contrast** drop-down list. This contrast compares (1) the overall mean number of correct responses obtained under the **Low Shock** and **High Shock** conditions (i.e., collapsing across the three types of items (Easy, Moderate, and Difficult)), (2) the overall mean number of correct responses obtained under the **Easy, Moderate,** and **Difficult** item conditions (i.e., collapsing across the two levels of shock (Low Shock and High Shock)), and (3) the overall mean number of correct responses obtained under the **Easy, Moderate,** and **Difficult** item conditions across the two levels of **SHOCK** (i.e., **SHOCK*ITEMS** interaction).

   Click [Continue] to return to the **Repeated Measures** window.

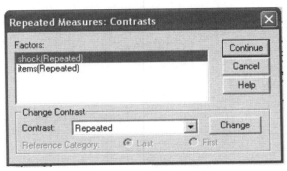

5. In the **Repeated Measures** window, click 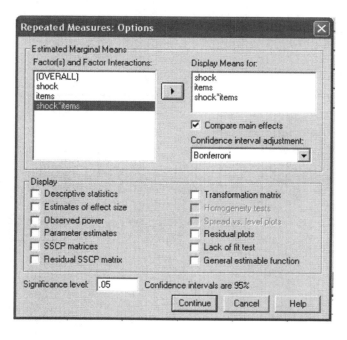 to open the **Repeated Measures: Options** window. To display means for the two levels of **SHOCK,** the three levels of **ITEMS,** and the **SHOCK*ITEMS** interaction, click (highlight) **SHOCK, ITEMS,** and **SHOCK*ITEMS** in the **Factor(s) and Factor Interactions** field, and

then click [▶] to transfer these variables to the **Display Means for** field. To compute pairwise comparisons of the means for the **SHOCK** (low shock vs. high shock), and **ITEMS** variables (easy vs. moderate, easy vs. difficult, moderate vs. difficult), check the **Compare main effects** field. However, because of inflated Type I error rate due to the multiple comparisons, a **Bonferroni** type adjustment should be made. From the **Confidence interval adjustment** field,

select **Bonferroni** from the drop-down list. Click [Continue] to return to the **Repeated Measures** window.

6. To aid interpretation of the **SHOCK*ITEMS** interaction effect, it would be useful to graph the means of this interaction. In the

Repeated Measures window, click [Plots...] to open the **Repeated Measures: Profile Plots** window. Transfer the **ITEMS** factor to the **Horizontal Axis** field by clicking it (highlighting) and then clicking

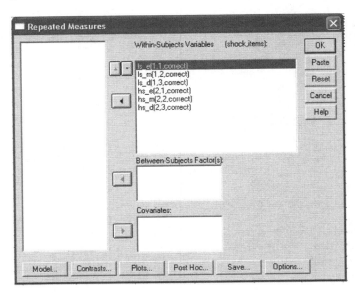 . Transfer the **SHOCK** factor to the **Separate Lines** field by clicking it (highlighting) and then clicking 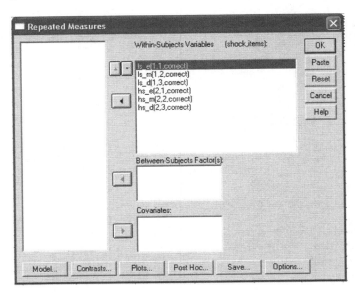 . Transfer this profile plot to the **Plots** field by clicking Add . Click Continue to return to the **Repeated Measures** window.

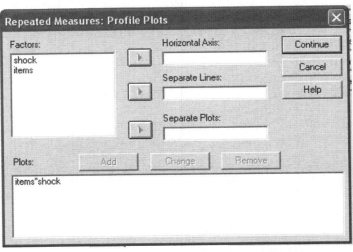

7. When the **Repeated Measures** window opens, click OK to complete the analysis. See Table 9.2 for the results.

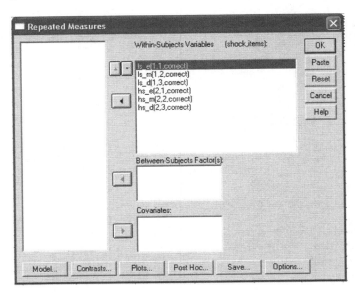

9.3.3 SPSS Syntax Method

GLM LS_E TO HS_D
/WSFACTOR=SHOCK 2 REPEATED ITEMS 3 REPEATED
/MEASURE=CORRECT
/PLOT=PROFILE(ITEMS*SHOCK)
/EMMEANS=TABLES(SHOCK) COMPARE ADJ(BONFERRONI)
/EMMEANS=TABLES(ITEMS) COMPARE ADJ(BONFERRONI)
/EMMEANS=TABLES(SHOCK*ITEMS).

9.3.4 SPSS Output

TABLE 9.2

Doubly Multivariate Repeated Measures Output

General Linear Model

Within-Subjects Factors

Measure: CORRECT

SHOCK	ITEMS	Dependent Variable
1	1	ls_e
	2	ls_m
	3	ls_d
2	1	hs_e
	2	hs_m
	3	hs_d

Multivariate Tests[b]

Effect		Value	F	Hypothesis df	Error df	Sig.
SHOCK	Pillai's Trace	.887	70.792[a]	1.000	9.000	.000
	Wilks' Lambda	.113	70.792[a]	1.000	9.000	.000
	Hotelling's Trace	7.866	70.792[a]	1.000	9.000	.000
	Roy's Largest Root	7.866	70.792[a]	1.000	9.000	.000
ITEMS	Piliai's Trace	.735	11.092[a]	2.000	8.000	.005
	Wilks' Lambda	.265	11.092[a]	2.000	8.000	.005
	Hotelling's Trace	2.773	11.092[a]	2.000	8.000	.005
	Roy's Largest Root	2.773	11.092[a]	2.000	8.000	.005
SHOCK * ITEMS	Pillars Trace	.745	11.684[a]	2.000	8.000	.004
	Wilks' Lambda	.255	11.684[a]	2.000	8.000	.004
	Hotelling's Trace	2.921	11.684[a]	2.000	8.000	.004
	Roy's Largest Root	2.921	11.684[a]	2.000	8.000	.004

[a]Exact statistic.
[b]Design: Intercept, Within Subjects Design: SHOCK + ITEMS + SHOCK*ITEMS.

Mauchly's Test of Sphericity[b]

Measure: CORRECT

Within Subjects Effect	Mauchly's W	Approx. Chi-Square	df	Sig.	Epsilon[a]		
					Greenhouse-Geisser	Huynh-Feldt	Lower-bound
SHOCK	1.000	.000	0	—	1.000	1.000	1.000
ITEMS	.372	7.913	2	.019	.614	.662	.500
SHOCK* ITEMS	.454	6.322	2	.042	.647	.709	.500

Note: Tests the null hypothesis that the error covariance matrix of the orthonormalized transformed dependent variables is proportional to an identity matrix.

[a]May be used to adjust the degrees of freedom for the averaged tests of significance. Corrected tests are displayed in the Tests of Within-Subjects Effects table.
[b]Design: Intercept, Within Subjects Design: SHOCK + ITEMS + SHOCK*ITEMS.

Tests of Within-Subjects Effects

Measure: CORRECT

Source		Type III Sum of Squares	df	Mean Square	F	Sig.
SHOCK	Sphericity Assumed	13380.267	1	13380.267	70.792	.000
	Greenhouse–Geisser	13380.267	1.000	13380.267	70.792	.000
	Huynh–Feldt	13380.267	1.000	13380.267	70.792	.000
	Lower-bound	13380.267	1.000	13380.267	70.792	.000
Error (SHOCK)	Sphericity Assumed	1701.067	9	189.007		
	Greenhouse–Gelsser	1701.067	9.000	189.007		
	Huynh–Feldt	1701.067	9.000	189.007		
	Lower-bound	1701.067	9.000	189.007		
ITEMS	Sphericity Assumed	1329.033	2	664.517	20.367	.000
	Greenhouse–Geisser	1329.033	1.228	1081.911	20.367	.001
	Huynh–Feldt	1329.033	1.323	1004.334	20.367	.000
	Lower-bound	1329.033	1.000	1329.033	20.367	.001
Error (ITEMS)	Sphericity Assumed	587.300	18	32.628		
	Greenhouse–Geisser	587.300	11.056	53.122		
	Huynh–Feldt	587.300	11.910	49.313		
	Lower-bound	587.300	9.000	65.256		
SHOCK * ITEMS	Sphericity Assumed	1023.433	2	511.717	12.683	.000
	Greenhouse–Geisser	1023.433	1.293	791.247	12.683	.003
	Huynh–Feldt	1023.433	1.419	721.313	12.683	.002
	Lower-bound	1023.433	1.000	1023.433	12.683	.006
Error (SHOCK * TEMS)	Sphericity Assumed	726.233	18	40.346		
	Greenhouse–Geisser	726.233	11.641	62.386		
	Huynh–Feldt	726.233	12.770	56.872		
	Lower-bound	726.233	9.000	80.693		

Tests of Within-Subjects Contrasts

Measure: CORRECT

Source	SHOCK	ITEMS	Type III Sum of Squares	df	Mean Square	F	Sig.
SHOCK	Level 1 vs. Level 2		8920.178	1	8920.178	70.792	.000
Error(SHOCK)	Level 1 vs. Level 2		1134.044	9	128.005		
ITEMS		Level 1 vs. Level 2	416.025	1	416.025	23.369	.001
		Level 2 vs. Level 3	255.025	1	255.025	11.757	.008
Error(ITEMS)		Level 1 vs. Level 2	160.225	9	17.803		
		Level 2 vs. Level 3	195.225	9	21.692		
SHOCK * ITEMS	Level 1 vs. Level 2	Level 1 vs. Level 2	828.100	1	828.100	18.969	.002
		Level 2 vs. Level 3	1232.100	1	1232.100	6.050	.036
Error(SHOCK * ITEMS)	Level 1 vs. Level 2	Level 1 vs. Level 2	392.900	9	43.656		
		Level 2 vs. Level 3	1832.900	9	203.656		

Tests of Between-Subjects Effects

Measure: CORRECT

Transformed Variable: Average

Source	Type III Sum of Squares	df	Mean Square	F	Sig.
Intercept	56801.344	1	56801.344	353.185	.000
Error	1447.433	9	160.826		

Estimated Marginal Means

1. SHOCK

Estimates

Measure: CORRECT

SHOCK	Mean	Standard Error	95% Confidence Interval Lower Bound	Upper Bound
1	90.300	4.097	81.033	99.567
2	60.433	4.657	49.899	70.967

Pairwise Comparisons

Measure: CORRECT

(I) SHOCK	(J) SHOCK	Mean Difference (I − J)	Standard Error	Sig.[a]	95% Confidence Interval for Difference[a]	
					Lower Bound	Upper Bound
1	2	29.867*	3.550	.000	21.837	37.897
2	1	−29.867*	3.550	.000	−37.897	−21.837

Note: Based on estimated marginal means.

*The mean difference is significant at the .05 level.

[a]Adjustment for multiple comparisons: Bonferroni.

Multivariate Tests

	Value	F	Hypothesis df	Error df	Sig.
Pillai's trace	.887	70.792[a]	1.000	9.000	.000
Wilks' lambda	.113	70.792[a]	1.000	9.000	.000
Hotelling's trace	7.866	70.792[a]	1.000	9.000	.000
Roy's largest root	7.866	70.792[a]	1.000	9.000	.000

Note: Each F tests the multivariate effect of SHOCK. These tests are based on the linearly independent pairwise comparisons among the estimated marginal means.

[a]Exact statistic.

2. ITEMS

Estimates

Measure: CORRECT

ITEMS	Mean	Std. Error	95% Confidence Interval	
			Lower Bound	Upper Bound
1	81.350	4.974	70.097	92.603
2	74.900	4.009	65.831	83.969
3	69.850	3.270	62.452	77.248

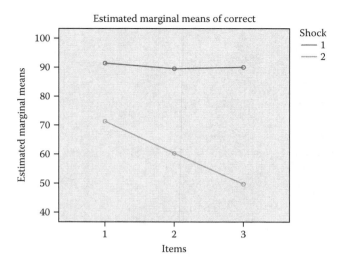

FIGURE 9.1

Pairwise Comparisons

Measure: CORRECT

(I) ITEMS	(J) ITEMS	Mean Difference (I − J)	Std. Error	Sig.a	95% Confidence Interval for Differencea	
					Lower Bound	Upper Bound
1	2	6.450*	1.334	.003	2.536	10.364
	3	11.500*	2.416	.003	4.412	18.588
2	1	−6.450*	1.334	.003	−10.364	−2.536
	3	5.050*	1.473	.023	.730	9.370
3	1	−11.500*	2.416	.003	−18.588	−4.412
	2	−5.050*	1.473	.023	−9.370	−.730

Note: Based on estimated marginal means.

*The mean difference is significant at the .05 level.

aAdjustment for multiple comparisons: Bonferroni.

Multivariate Tests

	Value	F	Hypothesis df	Error df	Sig.
Pillai's trace	.735	11.092a	2.000	8.000	.005
Wilks' lambda	.265	11.092a	2.000	8.000	.005
Hotelling's trace	2.773	11.092a	2.000	8.000	.005
Roy's largest root	2.773	11.092a	2.000	8.000	.005

Note: Each F tests the multivariate effect of ITEMS. These tests are based on the linearly independent pairwise comparisons among the estimated marginal means.

aExact statistic.

3. SHOCK*ITEMS

Measure: CORRECT

SHOCK	ITEMS	Mean	Std. Error	95% Confidence Interval Lower Bound	95% Confidence Interval Upper Bound
1	1	91.400	4.983	80.128	102.672
	2	89.500	3.481	81.626	97.374
	3	90.000	4.415	80.013	99.987
2	1	71.300	5.812	58.152	84.448
	2	60.300	5.243	48.439	72.161
	3	49.700	3.506	41.770	57.630

Profile Plots

9.3.5   Results and Interpretation

The **Multivariate Tests** involving the first within-subjects variable of **SHOCK** yielded highly significant results for all four multivariate tests (Pillai's, Hotelling's, Wilks', and Roy's), $F(1,9) = 70.79$, $p < .001$. From the **Estimated Marginal Means** table, it can be seen that on average, subjects obtained more correct responses under the low-shock condition ($M = 90.30$) than in the high-shock condition ($M = 60.43$).

For the second within-subjects variable of **ITEMS**, the **Mauchly's Test of Sphericity** yielded a value of 0.37 (Chi-square approximate value of 7.91 with 2 degrees of freedom) and is significant ($p < .05$). As the assumption of sphericity is violated, the **Multivariate Statistics** should be interpreted. (If the researcher chooses to interpret the **Tests of Within-Subjects Effects** statistics, then an adjustment must be made to the numerator and denominator degrees of freedom, using the Huynh–Feldt epsilon; the **Tests of Within-Subjects Effects** table displays the significance of each test when sphericity is assumed, or when the Huynh–Feldt epsilon is used.)

All four multivariate tests of significance (Pillai's, Hotelling's, Wilks', and Roy's) indicate that the within-subjects variable of **ITEMS** is significant, $F(2,8) = 11.09$, $p < .01$. From the **Estimated Marginal Means** table, it can be seen that on average, subjects obtained the most number of correct responses for the easy items ($M = 81.35$), less number of correct responses for the moderate items ($M = 74.90$), and the least number of correct responses for the difficult items ($M = 69.85$).

For the **SHOCK*ITEMS** interaction effect, the **Mauchly's Test of Sphericity** is significant, indicating that the assumption of sphericity is violated. Thus, the **Multivariate Tests** of significance will be interpreted. All four multivariate tests of significance indicate that the interaction effect is significant, $F(2,8) = 11.68$, $p < .01$, suggesting that the number of correct responses made across the three levels of difficulty of the items is dependent on the strength of the electrical shock. To aid interpretation of the interaction effect, it would be useful to examine the graph presented in Figure 9.1.

Figure 9.1 shows that the number of correct responses obtained across the three difficulty levels of the **ITEMS** variable is dependent on the strength of the electrical **SHOCK**. Under the low-shock condition (coded 1), subjects obtained similar number of correct responses across the three difficulty levels. However, under the high-shock condition (coded 2), the number of correct responses obtained decreased steadily from Easy to Moderate to Difficult.

The **Tests of Within-Subjects Contrasts** table can be used to compare the levels of the two within-subjects variables. For the **SHOCK** variable, the contrast (LEVEL 1 vs. LEVEL 2) compares the number of correct responses made under the low-shock condition (M = 90.30) with the number of correct responses made under the high-shock condition (M = 60.43), averaged across the three levels of **ITEMS** (see the estimated marginal means for SHOCK). This contrast is significant, $F(1,9) = 70.79$, $p < .001$.

For the **ITEMS** variable, the first contrast (LEVEL 1 vs. LEVEL 2) compares the number of correct responses made under the **easy items** condition (M = 81.35) with the number of correct responses made under the **moderate items** condition (M = 74.90), averaged across the two levels of **SHOCK**. This contrast is significant, $F(1,9) = 23.37$, $p < .01$. The second contrast (LEVEL 2 vs. LEVEL 3) compares the number of correct responses made under the **moderate items** condition (M = 74.9) with the number of correct responses made under the **difficult items** condition (M = 69.85), averaged across the two levels of **SHOCK** (see the estimated marginal means for ITEMS). This contrast is significant, $F(1,9) = 11.76$, $p < .01$.

The **SHOCK*ITEMS** contrast tests the hypothesis that the mean of the specified ITEMS contrast is the same across the two SHOCK levels. The first contrast (LEVEL 1 vs. LEVEL 2) is significant, $F(1,9) = 18.97$, $p < .01$, which indicates that the mean difference in the number of correct responses made between **EASY-ITEMS** and **MODERATE-ITEMS** is not the same across the two levels of SHOCK.

	Mean Difference (Easy Items vs. Moderate Items)
Low shock	1.90 (91.4 – 89.5)
High shock	11.00 (71.3 – 60.3)

In conjunction with Figure 9.1, the results indicate that the decrease in the number of correct responses made between **EASY-ITEMS** and **MODER-ATE-ITEMS** is different under the two shock conditions; the decrease is greater under the high-shock condition (11.00) than under the low-shock condition (1.90).

The second contrast (LEVEL 2 vs. LEVEL 3) is also significant, $F(1,9) = 6.05$, $p < .05$, which indicates that the mean difference in the number of correct responses made between **MODERATE-ITEMS** and **DIFFICULT-ITEMS** is not the same across the two levels of SHOCK.

	Mean Difference (Moderate Items vs. Difficult Items)
Low shock	−0.50 (89.50 − 90.00)
High shock	10.60 (60.30 − 49.70)

In conjunction with Figure 9.1, the results indicate that under the low-shock condition, there was an average increase of 0.50 correct responses, whereas under the high-shock condition, there was an average decrease of 10.60 correct responses.

The **Pairwise Comparisons** tables present all pairwise comparisons between the two levels of the **SHOCK** variable and the three levels of the **ITEMS** variable (with Bonferroni adjustment). For the **SHOCK** variable, the results indicate that the number of correct responses obtained under the low-shock condition (M = 90.30) is significantly greater ($p < .001$) than the number of correct responses obtained under the high-shock condition (M = 60.43). For the **ITEMS** variable, the results indicate that (1) the number of correct responses obtained under the easy condition (M = 81.35) is significantly greater ($p < .01$) than the number of correct responses obtained under the moderate (M = 74.90) and difficult (M = 69.85) conditions, and (2) the number of correct responses obtained under the moderate condition (M = 74.90) is significantly greater ($p < .05$) than the number of correct responses obtained under the difficult condition (M = 69.85).

9.4   Example 3 — GLM: Two-Factor Mixed Design (One Between-Groups Variable and One Within-Subjects Variable)

An experimenter wishes to determine the effects of positive reinforcement on rate of learning. The subjects are randomly assigned to three groups, one of which received "effort + ability" feedback, the second received "effort" feedback only, and the third group served as the control. All subjects were given three trials, and the number of correct responses per trial is recorded.

Group	Trial1	Trial2	Trial3
Control			
s1	3	3	4
s2	3	4	4
s3	1	3	4
s4	1	2	3
s5	2	3	5
s6	4	5	6
s7	4	5	6
s8	1	4	5
s9	1	4	5
s10	2	3	3
s11	2	2	4
Effort			
s12	2	3	6
s13	1	4	5
s14	3	6	7
s15	2	4	6
s16	1	2	3
s17	4	5	5
s18	1	3	4
s19	1	2	4
s20	2	3	4
s21	3	5	6
s22	4	6	7
Effort + Ability			
s23	2	4	8
s24	3	6	3
s25	4	7	7
s26	1	7	4
s27	3	7	12
s28	1	4	2
s29	1	5	3
s30	2	5	6
s31	3	6	6
s32	1	5	12
s33	4	9	7

9.4.1 Data Entry Format

The data set has been saved under the name: **EX9c.SAV**.

Variables	Column(s)	Code
Group	1	1 = control, 2 = effort, 3 = effort + ability
Trial1	2	Number of correct responses
Trial2	3	Number of correct responses
Trial3	4	Number of correct responses

9.4.2 Windows Method

1. From the menu bar, click **Analyze**, then **General Linear Model**, and then **Repeated Measures**. The following **Repeated Measures Define Factor(s)** window will open.

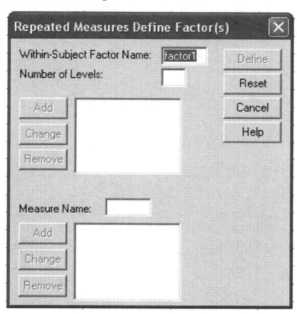

2. In the **Within-Subject Factor Name** field, type **TRIAL** to denote the name of the within-subject factor. In the **Number of Levels** field, type **3** to denote the three levels of the **TRIAL** factor (Trial1, Trial2, and Trial3). Click **Add** to transfer the **TRIAL** factor to the **ADD** field. In the **Measure Name** field, type the name of the dependent measure: **CORRECT**, and then click **Add** .

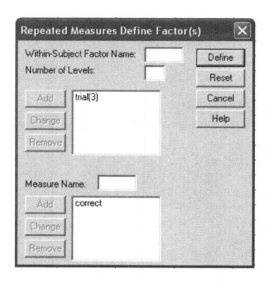

3. Click 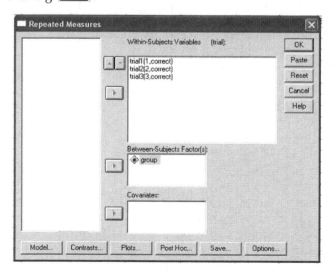 to open the **Repeated Measures** window. Transfer the three variables of Trial1, Trial2, and Trial3 to the **Within-Subjects Variables (trial)** field by clicking these variables (highlighting) and

then clicking ▶ . Next, transfer the variable of **GROUP** to the **Between-Subjects Factor(s)** field by clicking it (highlighting) and

then clicking ▶ .

4. To test for differences among the three trial levels (Trial1, Trial2, and Trial3), click <u>Contrasts...</u> to open the **Repeated Measures: Contrasts** window. Select **Repeated** as the contrast from the **Contrast** drop-down list. This contrast compares (1) the overall mean number of correct responses obtained under the Trial1, Trial2, and Trial3 conditions (i.e., collapsing across the three groups of Control, Effort, and Effort + Ability), and (2) the overall mean number of correct responses across the three trial conditions for the three groups (i.e., **TRIAL*GROUP** interaction). Select **None** as the contrast for the **GROUP** variable. Click <u>Continue</u> to return to the **Repeated Measures** window.

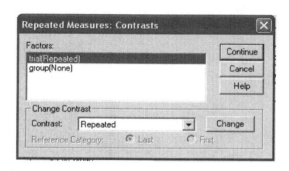

5. In the **Repeated Measures** window, click <u>Options...</u> to open the **Repeated Measures: Options** window. To display means for the three **GROUPS** (Control, Effort, and Effort + Ability), the three **TRIAL** levels (Trial1, Trial2, and Trial3), and the **GROUP*TRIAL** interaction levels, click (highlight) **GROUP, TRIAL,** and **GROUP*TRIAL** in the **Factor(s) and Factor Interactions** field, and then click <u>▶</u> to transfer these variables to the **Display Means for** field.

   To compute pairwise comparisons of the means for **TRIAL** (Trial1, Trial2, and Trial3), check the **Compare main effects** field. However, because of inflated Type I error rate due to the multiple comparisons, a **Bonferroni** type adjustment should be made. From the **Confidence interval adjustment** cell, select **Bonferroni** from the drop-down list.

   Click <u>Continue</u> to return to the **Repeated Measures** window.

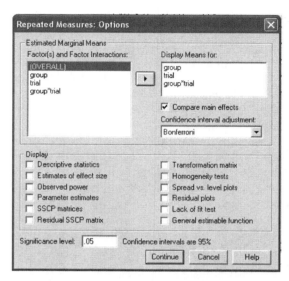

6. To test for differences between the three groups (Control, Effort, and Effort + Ability), conduct post hoc comparisons by clicking

Post Hoc... in the **Repeated Measures** window. This will open the **Repeated Measures: Post Hoc Multiple Comparisons for Observe** window. Transfer the **GROUP** variable to the **Post Hoc Tests for**

field by clicking it (highlighting) and then clicking ▶ . Check the **Scheffé** test field to run the Scheffé post hoc test. Next, click

Continue to return to the **Repeated Measures** window.

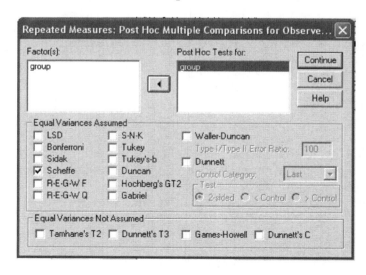

7. To aid interpretation of the **TRIAL*GROUP** interaction effect, it would be useful to graph the means of this interaction. In the **Repeated Measures** window, click [Plots...] to open the **Repeated Measures: Profile Plots** window. Transfer the **TRIAL** factor to the **Horizontal Axis** field by clicking it (highlighting) and then clicking [▶]. Transfer the **GROUP** factor to the **Separate Lines** field by clicking it (highlighting) and then clicking [▶]. Click [Add] to transfer this profile plot to the **Plots** field. Click [Continue] to return to the **Repeated Measures** window.

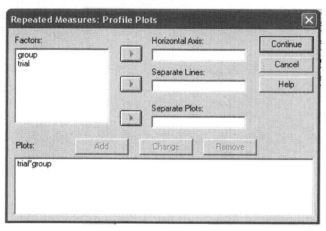

8. When the **Repeated Measures** window opens, click [OK] to complete the analysis. See Table 9.3 for the results.

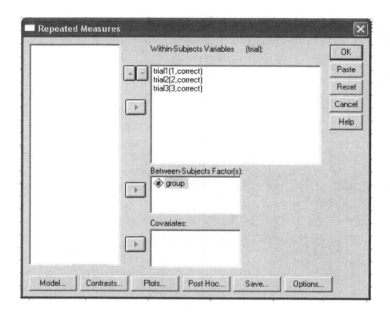

9.4.3   SPSS Syntax Method

GLM TRIAL1 TO TRIAL3 BY GROUP
/WSFACTOR=TRIAL 3 REPEATED
/MEASURE=CORRECT
/PLOT=PROFILE(TRIAL*GROUP)
/POSTHOC=GROUP(SCHEFFE)
/EMMEANS=TABLES(GROUP)
/EMMEANS=TABLES(TRIAL) COMPARE ADJ(BONFERRONI)
/EMMEANS=TABLES(GROUP*TRIAL).

9.4.4 SPSS Output

TABLE 9.3

Two-Factor Mixed Design Output

General Linear Model

Within-Subjects Factors

Measure: CORRECT

TRIAL	Dependent Variable
1	trial1
2	trial2
3	trial3

Between-Subjects Factors

Measure: CORRECT

		Value Label	N
group	1.00	control	11
	2.00	effort	11
	3.00	effort + ability	11

Multivariate Tests[c]

Effect		Value	F	Hypothesis df	Error df	Sig.
TRIAL	Pillai's Trace	.867	94.318[a]	2.000	29.000	.000
	Wilks' Lambda	.133	94.318[a]	2.000	29.000	.000
	Hotelling's Trace	6.505	94.318[a]	2.000	29.000	.000
	Roy's Largest Root	6.505	94.318[a]	2.000	29.000	.000
TRIAL * group	Pillai's Trace	.553	5.738	4.000	60.000	.001
	Wilks' Lambda	.451	7.099[a]	4.000	58.000	.000
	Hotelling's Trace	1.210	8.470	4.000	56.000	.000
	Roy's Largest Root	1.202	18.037[b]	2.000	30.000	.000

[a]Exact statistic.

[b]The statistic is an upper bound on F that yields a lower bound on the significance level.

[c]Design: Intercept + group, Within Subjects Design: TRIAL.

Mauchly's Test of Sphericity[b]

Measure: CORRECT

Within Subjects Effect	Mauchly's W	Approx. Chi-Square	df	Sig.	Greenhouse-Geisser	Huynh-Feldt	Lower-bound
					Epsilon[a]		
TRIAL	.507	19.712	2	.000	.670	.736	.500

Note: Tests the null hypothesis that the error covariance matrix of the orthonormalized transformed dependent variables is proportional to an identity matrix.

[a]May be used to adjust the degrees of freedom for the averaged tests of significance. Corrected tests are displayed in the Tests of Within-Subjects Effects table.

[b]Design: Intercept + group, Within Subjects Design: TRIAL.

Tests of within-Subjects Effects

Measure: CORRECT

Source		Type III Sum of Squares	df	Mean Square	F	Sig.
TRIAL	Sphericity Assumed	170.081	2	85.040	52.619	.000
	Greenhouse–Geisser	170.081	1.339	126.986	52.619	.000
	Huynh–Feldt	170.081	1.472	115.515	52.619	.000
	Lower-bound	170.081	1.000	170.081	52.619	.000
TRIAL * group	Sphericity Assumed	18.949	4	4.737	2.931	.028
	Greenhouse–Geisser	18.949	2.679	7.074	2.931	.050
	Huynh–Feldt	18.949	2.945	6.435	2.931	.045
	Lower-bound	18.949	2.000	9.475	2.931	.069
Error(TRIAL)	Sphericity Assumed	96.970	60	1.616		
	Greenhouse–Geisser	96.970	40.181	2.413		
	Huynh–Feldt	96.970	44.171	2.195		
	Lower-bound	96.970	30.000	3.232		

Tests of within-Subjects Contrasts

Measure: CORRECT

Source	TRIAL	Type III Sum of Squares	df	Mean Square	F	Sig.
TRIAL	Level 1 vs. Level 2	161.485	1	161.485	167.579	.000
	Level 2 vs. Level 3	27.273	1	27.273	6.347	.017
TRIAL * group	Level 1 vs. Level 2	34.606	2	17.303	17.956	.000
	Level 2 vs. Level 3	3.818	2	1.909	.444	.645
Error(TRIAL)	Level 1 vs. Level 2	28.909	30	.964		
	Level 2 vs. Level 3	128.909	30	4.297		

Tests of between-Subjects Effects

Measure: CORRECT
Transformed Variable: Average

Source	Type III Sum of Squares	df	Mean Square	F	Sig.
Intercept	525.337	1	525.337	333.102	.000
group	13.017	2	6.508	4.127	.026
Error	47.313	30	1.577		

Estimated Marginal Means

1. Group

Measure: CORRECT

Group	Mean	Standard Error	95% Confidence Interval Lower Bound	95% Confidence Interval Upper Bound
Control	3.364	.379	2.590	4.137
Effort	3.758	.379	2.984	4.531
Effort + ability	4.848	.379	4.075	5.622

2. Trial
Estimates

Measure: CORRECT

TRIAL	Mean	Standard Error	95% Confidence Interval Lower Bound	95% Confidence Interval Upper Bound
1	2.212	.205	1.794	2.630
2	4.424	.235	3.945	4.904
3	5.333	.380	4.557	6.109

Pairwise Comparisons

Measure: CORRECT

(I) TRIAL	(J) TRIAL	Mean Difference (I – J)	Std. Error	Sig.[a]	95% Confidence Interval for Difference[a] Lower Bound	Upper Bound
1	2	−2.212*	.171	.000	−2.645	−1.779
	3	−3.121*	.367	.000	−4.051	−2.191
2	1	2.212*	.171	.000	1.779	2.645
	3	−.909	.361	.052	−1.824	.006
3	1	3.121 *	.367	.000	2.191	4.051
	2	.909	.361	.052	−.006	1.824

Note: Based on estimated marginal means.

*The mean difference is significant at the .05 level.

[a]Adjustment for multiple comparisons: Bonferroni.

Multivariate Tests

	Value	F	Hypothesis df	Error df	Sig.
Pillai's trace	.867	94.318a	2.000	29.000	.000
Wilks' lambda	.133	94.318a	2.000	29.000	.000
Hotelling's trace	6.505	94.318a	2.000	29.000	.000
Roy's largest root	6.505	94.318a	2.000	29.000	.000

Each F tests the multivariate effect of TRIAL. These tests are based on the linearly independent pairwise comparisons among the estimated marginal means.

[a]Exact statistic.

3. Group*TRIAL

Measure: CORRECT

Group	TRIAL	Mean	Std. Error	95% Confidence Interval Lower Bound	Upper Bound
Control	1	2.182	.354	1.458	2.906
	2	3.455	.407	2.624	4.285
	3	4.455	.658	3.111	5.799
Effort	1	2.182	.354	1.458	2.906
	2	3.909	.407	3.079	4.739
	3	5.182	.658	3.838	6.526
Effort + ability	1	2.273	.354	1.549	2.997
	2	5.909	.407	5.079	6.739
	3	6.364	.658	5.020	7.708

Post Hoc Tests

Multiple Comparisons

Measure: CORRECT
Scheffe

		Mean Difference (I − J)	Standard Error	Sig.	95% Confidence Interval	
(I) group	(J) group				Lower Bound	Upper Bound
Control	Effort	−.3939	.53549	.765	−1.7729	.9850
	Effort + ability	−1.4848*	.53549	.033	−2.8638	−.1059
Effort	Control	.3939	.53549	.765	−.9850	1.7729
	Effort + ability	−1.0909	.53549	.143	−2.4699	.2881
Effort + ability	Control	1.4848*	.53549	.033	.1059	2.8638
	Effort	1.0909	.53549	.143	−.2881	2.4699

Note: Based on observed means.

*The mean difference is significant at the .05 level.

Homogeneous Subsets

CORRECT

Scheffe[a,b,c]

		Subset	
Group	N	1	2
Control	11	3.3636	
Effort	11	3.7576	3.7576
Effort + ability	11		4.8485
Sig.		.765	.143

Note: Means for groups in homogeneous subsets are displayed. Based on Type III Sum of Squares. The error term is Mean Square(Error) = 1.577.

[a]Uses Harmonic Mean Sample Size = 11.000.

[b]The group sizes are unequal. The harmonic mean of the group sizes is used. Type I error levels are not guaranteed.

[c]Alpha = .05.

Profile Plots

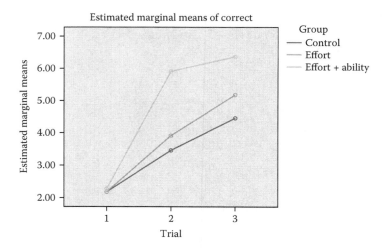

FIGURE 9.2

9.4.5   Results and Interpretation

9.4.5.1   Within-Subjects Effects

The **Multivariate Tests** test the within-subjects effects of **TRIAL** and **TRIAL*GROUP** interaction. The decision to interpret either the **Multivariate Tests** statistics or the **Tests of Within-Subjects Effects** depends on the outcome of the **Mauchly's Sphericity Test**. In this instance, the Mauchly's value is 0.507 (Chi-square approximate value of 19.71 with 2 degrees of freedom) and is significant ($p < 0.001$). Thus, the assumption of sphericity is violated, and the multivariate statistics should be interpreted. (Again, if the researcher chooses to interpret the **Tests of Within-Subjects Effects**, then an adjustment must be made to the numerator and denominator degrees of freedom using the Huynh–Feldt epsilon. The **Tests of Within-Subjects Effects** table displays the significance of each test when sphericity is assumed, or when the Huynh–Feldt epsilon is used.)

In the **Multivariate Tests** table, the main effect for the within-subjects variable of **TRIAL** is presented first, and is significant ($p < .001$), based on all four multivariate tests of significance (Pillai's, Wilks', Hotelling's, and Roy's). From the cell means presented in the **Estimated Marginal Means** table of **TRIAL**, the results indicate that the subjects made the least number of correct responses in Trial 1 (M = 2.21), more in Trial 2 (M = 4.42), and the most number of correct responses in Trial 3 (M = 5.33), averaged across the three groups.

The **Tests of Within-Subjects Contrasts** present the contrasts between the responses obtained across the three trials. The first contrast, LEVEL 1 vs.

LEVEL 2, compares the number of correct responses made in **Trial 1** (M = 2.21) with those made in **Trial 2** (M = 4.42), and is statistically significant, $F(1,30) = 167.58$, $p < .001$. The second contrast, LEVEL 2 vs. LEVEL 3, compares the number of correct responses made in **Trial 2** (M = 4.42) with the number of correct responses made in **Trial 3** (M = 5.33). The average value of this contrast is statistically significant, $F(1,30) = 6.35$, $p < .05$.

Whereas the **Tests of Within-Subjects Contrasts** test for differences between adjacent levels within the within-subjects variable of **TRIAL**, the **Pairwise Comparisons** (with Bonferroni correction) test for differences between *all* three levels. The comparison results presented in the **Pairwise Comparisons** table (under **Estimated Marginal Means**) indicate that the number of correct responses made in Trial 1 is significantly different (lower) from Trial 2 and Trial3 ($p < .001$). There is no significant difference between Trial 2 and Trial 3.

For the **TRIAL*GROUP** interaction, all four multivariate tests (Pillai's, Hotelling's, Wilks', and Roy's) indicate that this interaction is statistically significant ($p < .05$), suggesting that the number of correct responses made across the three trials is dependent on the type of treatment groups (i.e., Control, Effort, Effort + Ability). To aid interpretation of the interaction effect, it would be useful to examine the graph presented in Figure 9.2. Please note that the graph was plotted from the mean number of correct responses presented in the **Estimated Marginal Means** table of **GROUP*TRIAL** interaction.

Figure 9.2 shows that the subjects' performances across the three trials are dependent on the type of feedback they received. Although there is a general increase in the number of correct responses made across the three trials for all three groups, the rate of increase is greater for the "Effort + Ability" group from Trial 1 to Trial 2 than for the other two treatment groups.

The **Tests of Within-Subjects Contrasts** present the contrasts between the responses obtained across the three trials for the three groups. The first contrast, LEVEL 1 vs. LEVEL 2, is significant, $F(1,30) = 17.96$, $p < .001$, which indicates that the mean difference in the number of correct responses made between Trial 1 and Trial 2 is not the same for the three groups.

	Mean Difference (Trial 1 vs. Trial 2)
Control	−1.27 (2.18 − 3.45)
Effort	−1.73 (2.18 − 3.91)
Effort + Ability	−3.64 (2.27 − 5.91)

In conjunction with Figure 9.2, the results indicate that the increase in the number of correct responses made between Trial 1 and Trial 2 is different across the three groups; the increase is greatest for the Effort + Ability group (−3.64), less for the Effort group (−1.73), and least for the Control group (−1.27).

The second contrast, LEVEL 2 vs. LEVEL 3, is not significant ($p > .05$), which indicates that the mean difference in the number of correct responses made in Trial 2 and Trial 3 is the same for the three groups.

	Mean Difference (Trial 2 vs. Trial 3)
Control	−1.00 (3.45 − 4.45)
Effort	−1.27 (3.91 − 5.18)
Effort + Ability	−0.45 (5.91 − 6.36)

In conjunction with Figure 9.2, the results indicate that the increase in the number of correct responses made between Trial 2 and Trial 3 is similar across the three groups.

9.4.5.2   Between-Groups Effect

The **Tests of Between-Subjects Effects** is equivalent to a one-way analysis of variance. The results indicate that the between-groups variable of **GROUP** is statistically significant, $F(2,30) = 4.13$, $p < .05$. From the cell means presented in the **Estimated Marginal Means** table of **GROUP**, the results indicate that the Control group made the least number of correct responses (M = 3.36), with the Effort group making more correct responses (M = 3.76), and the Effort + Ability group making the most number of correct responses (M = 4.85), averaged across the three trials.

The post hoc Scheffé multiple comparisons indicate that only the Control group (M = 3.36) and the Effort + Ability group (M = 4.85) differed significantly ($p < .05$), averaged across the three trials.

9.5   Example 4 — GLM: Three-Factor Mixed Design (Two Between-Groups Variables and One Within-Subjects Variable)

In addition to investigating the effect of a drug assumed to increase learning ability, a researcher also wanted to investigate the effect of subject's gender on performance over a series of trials. This study incorporates (1) the between-groups variables of drug-present or drug-absent, and male vs. female subjects, and (2) the within-subjects variables of trials. Twenty subjects have been randomly assigned to the four groups. The number of errors per trial is recorded.

			Trials			
		Subjects	T1	T2	T3	T4
Drug present	Male	s1	14	10	10	8
		s2	16	9	10	5
		s3	18	16	12	7
		s4	17	14	11	8
		s5	22	19	17	10
	Female	s6	18	19	17	15
		s7	25	24	20	16
		s8	37	35	32	24
		s9	21	18	15	14
		s10	29	29	28	26
Drug absent	Male	s11	17	15	10	7
		s12	24	12	9	9
		s13	28	20	14	12
		s14	30	22	16	12
		s15	22	16	9	4
	Female	s16	42	34	30	22
		s17	29	28	26	22
		s18	26	25	20	16
		s19	30	30	25	14
		s20	26	27	20	12

9.5.1 Data Entry Format

The data set has been saved under the name: **EX9d.SAV**.

Variables	Column(s)	Code
Drug	1	1 = present, 2 = absent
Sex	2	1 = male, 2 = female
Trial1	3	Number of errors
Trial2	4	Number of errors
Trial3	5	Number of errors
Trial4	6	Number of errors

9.5.2 Windows Method

1. From the menu bar, click **Analyze**, then **General Linear Model**, and then **Repeated Measures**. The following **Repeated Measures Define Factor(s)** window will open:

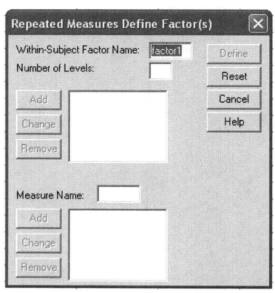

2. In the **Within-Subject Factor Name** field, type **TRIAL** to denote the name of the within-subject factor. In the **Number of Levels** field, type **4** to denote the four levels of the **TRIAL** factor (Trial1, Trial2, Trial3, and Trial4). Click Add to transfer the **TRIAL** factor to the **ADD** field. In the **Measure Name** field, type the name of the dependent measure: **ERRORS**, and then click Add .

3. Click 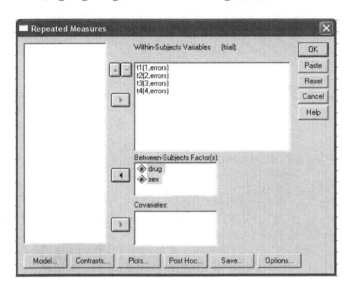 **Define** to open the **Repeated Measures** window. Transfer the four variables of Trial1, Trial2, Trial3, and Trial4 to the **Within-Subjects Variables (trial)** field by clicking these variables (highlighting) and then clicking ▶ . Next, transfer the **DRUG** and **SEX** variables to the **Between-Subjects Factor(s)** field by clicking these variables (highlighting) and then clicking ▶ .

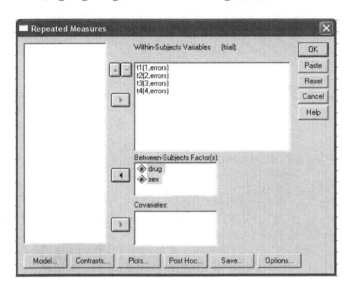

4. To test for differences among the four trial levels (Trial1, Trial2, Trial3, and Trial4), click **Contrasts...** to open the **Repeated Measures: Contrasts** window. Select **Repeated** as the contrast from the **Contrast** drop-down list. This contrast compares (1) the overall mean number of error responses obtained under the Trial1, Trial2, Trial3, and Trial4 conditions (i.e., collapsing across the **DRUG** and **SEX** variables), (2) the overall mean number of error responses across the four trial conditions for the two **DRUG** groups (i.e., **TRIAL*DRUG** interaction collapsing across the **SEX** variable), (3) the overall mean number of error responses across the four trial conditions for the two **SEX** groups (i.e., **TRIAL*SEX** interaction collapsing across the **DRUG** variable), and (4) the overall mean number of error responses across the four trial conditions as a function of the **DRUG*SEX** interaction (i.e., **TRIAL*DRUG*SEX** interaction). Select **None** as the contrast for the **DRUG** and **SEX** variables.

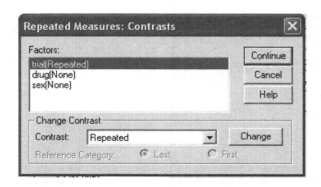

Click to return to the **Repeated Measures** window.

5. In the **Repeated Measures** window, click **Options...** to open the **Repeated Measures: Options** window. To display the mean number of errors for the two groups of **DRUG**, the two groups of **SEX**, the four **TRIAL** levels (Trial1, Trial2, Trial3, Trial4), the **DRUG*SEX** interaction, the **DRUG*TRIAL** interaction, the **SEX*TRIAL** interaction, and the **DRUG*SEX*TRIAL** interaction, click (highlight) **DRUG, SEX, TRIAL, DRUG*SEX, DRUG*TRIAL, SEX*TRIAL, DRUG*SEX*TRIAL** in the **Factor(s) and Factor Interactions** field, and then click ▶ to transfer these variables to the **Display Means for** field.

To compute pairwise comparisons of the means for **TRIAL** (Trial1, Trial2, Trial3, and Trial4), check the **Compare main effects** field. However, because of inflated Type I error rate due to the multiple comparisons, a **Bonferroni** type adjustment should be made. From the **Confidence interval adjustment** cell, select **Bonferroni** from the drop-down list. Click **Continue** to return to the **Repeated Measures** window.

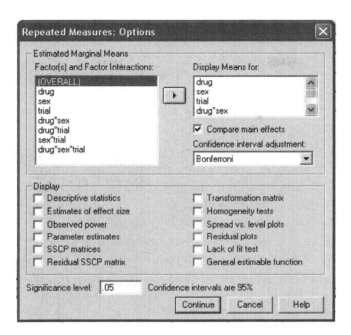

6. To aid interpretation of the **DRUG*SEX, DRUG*TRIAL**, and **SEX*TRIAL** interaction effects, it would be useful to graph the means of these interactions. In the **Repeated Measures** window, click [Plots...] to open the **Repeated Measures: Profile Plots** window. For the **DRUG*SEX** interaction, transfer the **SEX** factor to the **Horizontal Axis** field by clicking it (highlighting) and then clicking [▶]. Transfer the **DRUG** factor to the **Separate Lines** field by clicking it (highlighting) and then clicking [▶]. Click [Add] to transfer this profile plot to the **Plots** field. For the **TRIAL*DRUG** interaction, transfer the **TRIAL** factor to the **Horizontal Axis** field by clicking it (highlighting) and then clicking [▶]. Transfer the **DRUG** factor to the **Separate Lines** field by clicking it (highlighting) and then clicking [▶]. Click [Add] to transfer this profile plot to the **Plots** field. For the **TRIAL*SEX** interaction, transfer the **TRIAL** factor to the **Horizontal Axis** field by clicking it (highlighting) and then clicking [▶]. Transfer the **SEX** factor to the **Separate**

Lines field by clicking it (highlight) and then clicking [▶] . Click
[Add] to transfer this profile plot to the **Plots** field. Click
[Continue] to return to the **Repeated Measures** window.

7. To interpret the full three-way interaction (**TRIAL*DRUG*SEX**), this interaction should also be graphed. This three-way interaction entails plotting the within-subjects factor of **TRIAL** against the factorial combination of the two between-groups factors of **DRUG** and **SEX,** i.e., against the four groups of **drug-present male, drug-present female, drug-absent male**, and **drug-absent female**. To do this, some data transformation must be carried out.

7a. Data transformation

7a.1 The first step is to create a new grouping variable called **GROUP** that contains the four levels generated by the **DRUG*SEX** interaction (**drug-present male, drug-present female, drug-absent male, drug-absent female**). From the menu bar, click **Transform** and then **Compute** to open the **Compute Variable** window.

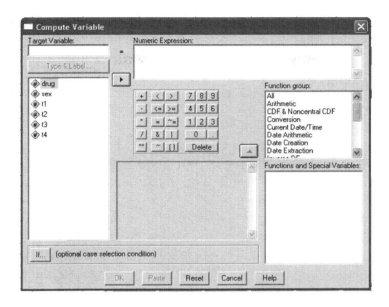

7a.2 Click 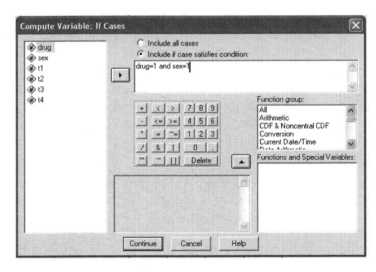 to open the **Compute Variable: If Cases** window. Ensure that the **Include if case satisfies condition** field is checked. To create the first level (**drug-present male**), type

drug=1 and sex=1 in the field. Click **Continue**.

7a.3 Because the **drug-present male** level is the first of four levels within a new variable called **GROUP**, this level will be coded

1 within the **GROUP** variable. When the **Compute Variable** window opens, type **GROUP** in the **Target Variable** field and

1 in the **Numeric Expression** cell. Click [OK] to create the first level of **drug-present male** (coded 1) within the new grouping variable of **GROUP.**

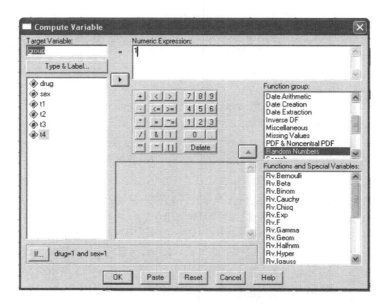

7a.4 Repeat step 7a.1 to step 7a.3 to create the other three levels: **drug-present female** (coded 2), **drug-absent male** (coded 3), and **drug-absent female** (coded 4).

7a.5 To aid interpretation of the obtained results, **Value Labels** in the data set should be activated and labels attached to the numerical codes for the four levels. To do this, open the data set, and under **Variable View**, click the **Values** cell for the newly created **GROUP** variable. Type the value labels as indicated in the **Value Labels** window. Click [OK] .

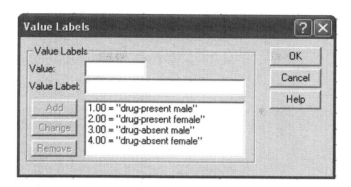

8. Open the **Repeated Measures** window by repeating step 1 to step 3. In the **Repeated Measures** window, transfer the **GROUP** variable to the **Between-Subjects Factor(s)** field.

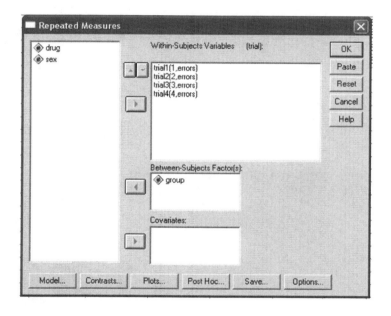

9. Click **Plots...** to open the **Repeated Measures: Profile Plots** window. To graph the full three-way **DRUG*SEX*TRIAL** interaction, transfer the **TRIAL** factor to the **Horizontal Axis** field by clicking it (highlighting) and then clicking ▶. Transfer the newly created **GROUP** factor to the **Separate Lines** field by clicking it

(highlighting) and then clicking ▶ . Click **Add** to transfer this profile plot to the **Plots** field. Click **Continue** to return to the **Repeated Measures** window.

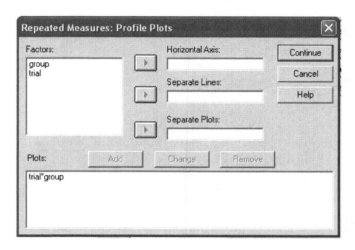

10. When the **Repeated Measures** window opens, click **OK** to complete the analysis. See Table 9.4 for the results.

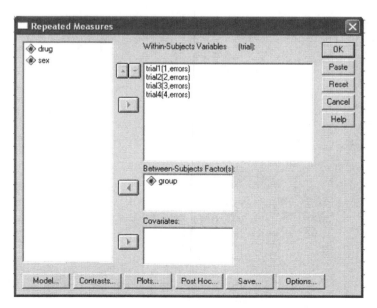

9.5.3   SPSS Syntax Method

GLM TRIAL1 TO TRIAL4 BY DRUG SEX
/WSFACTOR=TRIAL 4 REPEATED
/MEASURE=ERRORS
/EMMEANS=TABLES(DRUG)
/EMMEANS=TABLES(SEX)
/EMMEANS=TABLES(DRUG*SEX)
/EMMEANS=TABLES(TRIAL) COMPARE ADJ(BONFERRONI)
/EMMEANS=TABLES(DRUG*TRIAL)
/EMMEANS=TABLES(SEX*TRIAL)
/EMMEANS=TABLES(DRUG*SEX*TRIAL)
/PLOT=PROFILE(DRUG*SEX)
/PLOT=PROFILE(TRIAL*DRUG)
/PLOT=PROFILE(TRIAL*SEX).

IF (DRUG EQ 1 AND SEX EQ 1) GROUP=1.
IF (DRUG EQ 2 AND SEX EQ 1) GROUP=2.
IF (DRUG EQ 1 AND SEX EQ 2) GROUP=3.
IF (DRUG EQ 2 AND SEX EQ 2) GROUP=4.

VALUE LABELS GROUP   1 'DRUG PRESENT-MALE'
                         2 'DRUG ABSENT-MALE'
                         3 'DRUG PRESENT-FEMALE'
                         4 'DRUG ABSENT-FEMALE'.

GLM TRIAL1 TO TRIAL4 BY GROUP
/WSFACTOR=TRIAL 4 REPEATED
/MEASURE=ERRORS
/PLOT=PROFILE(TRIAL*GROUP)
/EMMEANS=TABLES(GROUP*TRIAL).

9.5.4   SPSS Output

TABLE 9.4

Three-Factor Mixed Design Output

General Linear Model

Within-Subjects Factors

Measure: ERRORS

TRIAL	Dependent Variable
1	triall
2	tria12
3	tria13
4	trial4

Between-Subjects Factors

	Value	Label	N
Drug	1.00	present	10
	2.00	absent	10
Sex	1.00	male	10
	2.00	female	10

Multivariate Test[b]

Effect		Value	F	Hypothesis df	Error df	Sig.
TRIAL	Pillars Trace	.927	59.694[a]	3.000	14.000	.000
	Wilks' Lambda	.073	59.694[a]	3.000	14.000	.000
	Hotelling's Trace	12.792	59.694[a]	3.000	14.000	.000
	Roys Largest Root	12.792	59.694[a]	3.000	14.000	.000
TRIAL * drug	Pillai's Trace	.639	8.259[a]	3.000	14.000	.002
	Wilks' Lambda	.361	8.259[a]	3.000	14.000	.002
	Hotelling's Trace	1.770	8.259[a]	3.000	14.000	.002
	Roys Largest Root	1.770	8.259[a]	3.000	14.000	.002
TRIAL * sex	Pillai's Trace	.488	4.448[a]	3.000	14.000	.021
	Wilks' Lambda	.512	4.448[a]	3.000	14.000	.021
	Hotelling's Trace	.953	4.448[a]	3.000	14.000	.021
	Roy's Largest Root	.953	4.448[a]	3.000	14.000	.021
TRIAL * drug * sex	Pillai's Trace	.483	4.366[a]	3.000	14.000	.023
	Wilks' Lambda	.517	4.366[a]	3.000	14.000	.023
	Hotelling's Trace	.936	4.366[a]	3.000	14.000	.023
	Roys Largest Root	.936	4.366[a]	3.000	14.000	.023

[a]Exact statistic.

[b]Design: Intercept + drug + sex + drug * sex, Within Subjects Design: TRIAL.

Mauchly's Test of Sphericity[b]

Measure: ERRORS

Within Subjects Effect	Mauchly's W	Approx. Chi-Square	df	Sig.	Epsilon[a]		
					Greenhouse-Geisser	Huynh-Feldt	Lower-bound
TRIAL	.298	17.816	5	.003	.674	.917	.333

Tests the null hypothesis that the error covariance matrix of the orthonormalized transformed dependent variables is proportional to an Identity matrix.

[a]May be used to adjust the degrees of freedom for the averaged tests of significance. Corrected tests are displayed in the Tests of Within-Subjects Effects table.
[b]Design: Intercept + drug + sex + drug * sex, Within Subjects Design: TRIAL.

Tests of within-Subjects Effects

Measure: ERRORS

Source		Type III Sum of Squares	df	Mean Square	F	Sig.
TRIAL	Sphericity Assumed	1430.137	3	476.712	111.295	.000
	Greenhouse–Geisser	1430.137	2.022	707.140	111.295	.000
	Huynh–Feldt	1430.137	2.751	519.912	111.295	.000
	Lower-bound	1430.137	1.000	1430.137	111.295	.000
TRIAL * drug	Sphericity Assumed	112.538	3	37.513	8.758	.000
	Greenhouse–Geisser	112.538	2.022	55.645	8.758	.001
	Huynh–Feldt	112.538	2.751	40.912	8.758	.000
	Lower-bound	112.538	1.000	112.538	8.758	.009
TRIAL * sex	Sphericity Assumed	55.038	3	18.346	4.283	.009
	Greenhouse–Geisser	55.038	2.022	27.214	4.283	.022
	Huynh–Feldt	55.038	2.751	20.008	4.283	.012
	Lower-bound	55.038	1.000	55.038	4.283	.055
TRIAL* drug * sex	Sphericity Assumed	21.438	3	7.146	1.668	.186
	Greenhouse–Geisser	21.438	2.022	10.600	1.668	.204
	Huynh–Feldt	21.438	2.751	7.793	1.668	.191
	Lower-bound	21.438	1.000	21.438	1.668	.215
Error(TRIAL)	Sphericity Assumed	205.600	48	4.283		
	Greenhouse–Geisser	205.600	32.359	6.354		
	Huynh–Feldt	205.600	44.012	4.671		
	Lower-bound	205.600	16.000	12.850		

Tests of within-Subjects Contrasts

Measure: ERRORS

Source	TRIAL	Type III Sum of Squares	df	Mean Square	F	Sig.
TRIAL	Level 1 vs. Level 2	238.050	1	238.050	29.664	.000
	Level 2 vs. Level 3	252.050	1	252.050	90.018	.000
	Level 3 vs. Level 4	387.200	1	387.200	63.737	.000
TRIAL * drug	Level 1 vs. Level 2	22.050	1	22.050	2.748	.117
	Level 2 vs. Level 3	42.050	1	42.050	15.018	.001
	Level 3 vs. Level 4	5.000	1	5.000	.823	.378
TRIAL * sex	Level 1 vs. Level 2	84.050	1	84.050	10.474	.005
	Level 2 vs. Level 3	.050	1	.050	.018	.895
	Level 3 vs. Level 4	12.800	1	12.800	2.107	.166
TRIAL* drug * sex	Level 1 vs. Level 2	8.450	1	8.450	1.053	.320
	Level 2 vs. Level 3	4.050	1	4.050	1.446	.247
	Level 3 vs. Level 4	33.800	1	33.800	5.564	.031
Error(TRIAL)	Level 1 vs. Level 2	128.400	16	8.025		
	Level 2 vs. Level 3	44.800	16	2.800		
	Level 3 vs. Level 4	97.200	16	6.075		

Tests of between-Subjects Effects

Measure: ERRORS

Transformed Variable: Average

Source	Type III Sum of Squares	df	Mean Square	F	Sig.
Intercept	7286.653	1	7286.653	347.604	.000
Drug	29.403	1	29.403	1.403	.254
Sex	512.578	1	512.578	24.452	.000
Drug * sex	.528	1	.528	.025	.876
Error	335.400	16	20.963		

Estimated Marginal Means

1. Drug

Measure: ERRORS

Drug	Mean	Std. Error	95% Confidence Interval Lower Bound	95% Confidence Interval Upper Bound
Present	17.875	1.448	14.806	20.944
Absent	20.300	1.448	17.231	23.369

2. Sex

Measure: ERRORS

Sex	Mean	Std. Error	95% Confidence Interval	
			Lower Bound	Upper Bound
Male	14.025	1.448	10.956	17.094
Female	24.150	1.448	21.081	27.219

3. Drug*sex

Measure: ERRORS

Drug	Sex	Mean	Std. Error	95% Confidence Interval	
				Lower Bound	Upper Bound
Present	Male	12.650	2.048	8.309	16.991
	Female	23.100	2.048	18.759	27.441
Absent	Male	15.400	2.048	11.059	19.741
	Female	25.200	2.048	20.859	29.541

4. Trial
Estimates

Measure: ERRORS

TRIAL	Mean	Std. Error	95% Confidence Interval	
			Lower Bound	Upper Bound
1	24.550	1.293	21.808	27.292
2	21.100	1.093	18.784	23.416
3	17.550	1.062	15.298	19.802
4	13.150	.917	11.207	15.093

Pairwise Comparisons

Measure: ERRORS

(I) TRIAL	(J) TRIAL	Mean Difference (I − J)	Std. Error	Sig.[a]	95% Confidence Interval for Difference[a]	
					Lower Bound	Upper Bound
1	2	3.450*	.633	.000	1.544	5.356
	3	7.000*	.594	.000	5.214	8.786
	4	11.400*	.868	.000	8.788	14.012
2	1	−3.450*	.633	.000	−5.356	−1.544
	3	3.550*	.374	.000	2.424	4.676
	4	7.950*	.787	.000	5.584	10.316
3	1	−7.000*	.594	.000	−8.786	−5.214
	2	−3.550*	.374	.000	−4.676	−2.424
	4	4.400*	.551	.000	2.742	6.058
4	1	−11.400*	.868	.000	−14.012	−8.788
	2	−7.950*	.787	.000	−10.316	−5.584
	3	−4.400*	.551	.000	−6.058	−2.742

Note: Based on estimated marginal means.

*The mean difference is significant at the .05 level.

[a]Adjustment for multiple comparisons: Bonferroni.

Multivariate Tests

	Value	F	Hypothesis df	Error df	Sig.
Pillai's trace	.927	59.694[a]	3.000	14.000	.000
Wilks' lambda	.073	59.694[a]	3.000	14.000	.000
Hotelling's trace	12.792	59.694[a]	3.000	14.000	.000
Roy's largest root	12.792	59.694[a]	3.000	14.000	.000

Note: Each F tests the multivariate effect of TRIAL. These tests are based on the linearly independent pairwise comparisons among the estimated marginal means.

[a]Exact statistic.

5. drug * TRIAL

Measure: ERRORS

Drug	TRIAL	Mean	Std. Error	95% Confidence Interval	
				Lower Bound	Upper Bound
Present	1	21.700	1.829	17.823	25.577
	2	19.300	1.545	16.024	22.576
	3	17.200	1.502	14.015	20.385
	4	13.300	1.296	10.552	16.048
Absent	1	27.400	1.829	23.523	31.277
	2	22.900	1.545	19.624	26.176
	3	17.900	1.502	14.715	21.085
	4	13.000	1.296	10.252	15.748

6. Sex * TRIAL

Measure: ERRORS

Sex	TRIAL	Mean	Std. Error	95% Confidence Interval Lower Bound	95% Confidence Interval Upper Bound
Male	1	20.800	1.829	16.923	24.677
	2	15.300	1.545	12.024	18.576
	3	11.800	1.502	8.615	14.985
	4	8.200	1.296	5.452	10.948
Female	1	28.300	1.829	24.423	32.177
	2	26.900	1.545	23.624	30.176
	3	23.300	1.502	20.115	26.485
	4	18.100	1.296	15.352	20.848

7. Drug * sex * TRIAL

Measure: ERRORS

Drug	Sex	TRIAL	Mean	Std. Error	95% Confidence Interval Lower Bound	95% Confidence Interval Upper Bound
Present	Male	1	17.400	2.587	11.917	22.883
		2	13.600	2.185	8.968	18.232
		3	12.000	2.125	7.496	16.504
		4	7.600	1.833	3.714	11.486
	Female	1	26.000	2.587	20.517	31.483
		2	25.000	2.185	20.368	29.632
		3	22.400	2.125	17.896	26.904
		4	19.000	1.833	15.114	22.886
Absent	Male	1	24.200	2.587	18.717	29.683
		2	17.000	2.185	12.368	21.632
		3	11.600	2.125	7.096	16.104
		4	8.800	1.833	4.914	12.686
	Female	1	30.600	2.587	25.117	36.083
		2	28.800	2.185	24.168	33.432
		3	24.200	2.125	19.696	28.704
		4	17.200	1.833	13.314	21.086

Profile Plots

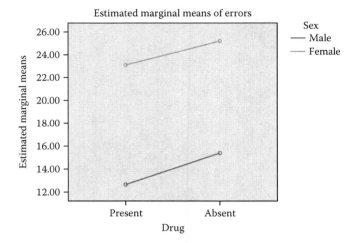

FIGURE 9.3

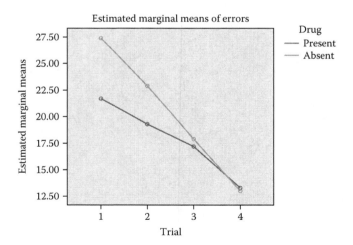

FIGURE 9.4

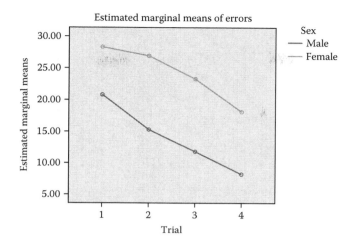

FIGURE 9.5

General Linear Model

Within-Subjects Factors

Measure: ERRORS

TRIAL	Dependent Variable
1	Trial1
2	Trial2
3	Trial3
4	Trial4

Between-Subjects Factors

		Value Label	N
GROUP	1.00	DRUG PRESENT-MALE	5
	2.00	DRUG ABSENT-MALE	5
	3.00	DRUG PRESENT-FEMALE	5
	4.00	DRUG ABSENT- FEMALE	5

Multivariate Tests[c]

Effect		Value	F	Hypothesis df	Error df	Sig.
TRIAL	Pillai's Trace	.927	59.694[a]	3.000	14.000	.000
	Wilks' Lambda	.073	59.694[a]	3.000	14.000	.000
	Hotelling's Trace	12.792	59.694[a]	3.000	14.000	.000
	Roy's Largest Root	12.792	59.694[a]	3.000	14.000	.000
TRIAL * GROUP	Pillai's Trace	1.267	3.901	9.000	48.000	.001
	Wilks' Lambda	.138	4.785	9.000	34.223	.000
	Hotelling's Trace	3.659	5.149	9.000	38.000	.000
	Roy's Largest Root	2.924	15.595[b]	3.000	16.000	.000

[a]Exact statistic.
[b]The statistic is an upper bound on F that yields a lower bound on the significance level.
[c]Design: Intercept + GROUP, Within-Subjects Design: TRIAL.

Mauchly's Test of Sphericity[b]

Measure: ERRORS

Within Subjects Effect	Mauchly's W	Approx. Chi-Square	df	Sig.	Epsilon[a] Greenhouse-Geisser	Huynh-Feldt	Lower-bound
TRIAL	.298	17.816	5	.003	.674	.917	.333

Note: Tests the null hypothesis that the error covariance matrix of the orthonormalized transformed dependent variables is proportional to an identity matrix.

[a]May be used to adjust the degrees of freedom for the averaged tests of significance. Corrected tests are displayed in the Tests of within-Subjects Effects table.
[b]Design: Intercept + GROUP, Within-Subjects Design: TRIAL.

Tests of within-Subjects Effects

Measure: ERRORS

Source		Type III Sum of Squares	df	Mean Square	F	Sig.
TRIAL	Sphericity Assumed	1430.137	3	476.712	111.295	.000
	Greenhouse–Geisser	1430.137	2.022	707.140	111.295	.000
	Huynh–Feldt	1430.137	2.751	519.912	111.295	.000
	Lower-bound	1430.137	1.000	1430.137	111.295	.000
TRIAL * GROUP	Sphericity Assumed	189.013	9	21.001	4.903	.000
	Greenhouse–Geisser	189.013	6.067	31.153	4.903	.001
	Huynh–Feldt	189.013	8.252	22.905	4.903	.000
	Lower-bound	189.013	3.000	63.004	4.903	.013
Error(TRIAL)	Sphericity Assumed	205.600	48	4.283		
	Greenhouse–Geisser	205.600	32.359	6.354		
	Huynh–Feldt	205.600	44.012	4.671		
	Lower-bound	205.600	16.000	12.850		

Tests of within-Subjects Contrasts

Measure: ERRORS

Source	TRIAL	Type III Sum of Squares	df	Mean Square	F	Sig.
TRIAL	Level 1 vs. Level 2	238.050	1	238.050	29.664	.000
	Level 2 vs. Level 3	252.050	1	252.050	90.018	.000
	Level 3 vs. Level 4	387.200	1	387.200	63.737	.000
TRIAL * GROUP	Level 1 vs. Level 2	114.550	3	38.183	4.758	.015
	Level 2 vs. Level 3	46.150	3	15.383	5.494	.009
	Level 3 vs. Level 4	51.600	3	17.200	2.831	.071
Error(TRIAL)	Level 1 vs. Level 2	128.400	16	8.025		
	Level 2 vs. Level 3	44.800	16	2.800		
	Level 3 vs. Level 4	97.200	16	6.075		

Tests of between-Subjects Effects

Measure: ERRORS

Transformed Variable: Average

Source	Type III Sum of Squares	df	Mean Square	F	Sig.
Intercept	7286.653	1	7286.653	347.604	.000
Group	542.509	3	180.836	8.627	.001
Error	335.400	16	20.963		

Estimated Marginal Means

GROUP * TRIAL

Measure: ERRORS

GROUP	TRIAL	Mean	Std. Error	95% Confidence Interval Lower Bound	Upper Bound
DRUG PRESENT-MALE	1	17.400	2.587	11.917	22.883
	2	13.600	2.185	8.968	18.232
	3	12.000	2.125	7.496	16.504
	4	7.600	1.833	3.714	11.486
DRUG ABSENT-MALE	1	24.200	2.587	18.717	29.683
	2	17.000	2.185	12.368	21.632
	3	11.600	2.125	7.096	16.104
	4	8.800	1.833	4.914	12.686
DRUG PRESENT-FEMALE	1	26.000	2.587	20.517	31.483
	2	25.000	2.185	20.368	29.632
	3	22.400	2.125	17.896	26.904
	4	19.000	1.833	15.114	22.886
DRUG ABSENT-FEMALE	1	30.600	2.587	25.117	36.083
	2	28.800	2.185	24.168	33.432
	3	24.200	2.125	19.696	28.704
	4	17.200	1.833	13.314	21.086

Profile Plots

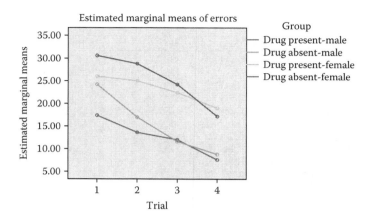

FIGURE 9.6

9.5.5   Results and Interpretation

9.5.5.1   *Within-Subjects Effects*

As the **Mauchly's Test of Sphericity** is statistically significant ($p < .01$), the assumption of sphericity is violated, and the multivariate statistics should be interpreted. The **Multivariate Tests** test the within-subjects effects of **TRIAL, TRIAL*DRUG, TRIAL*SEX**, and **TRIAL*DRUG*SEX** interactions.

9.5.5.1.1   *TRIAL Main Effect*

In the **Multivariate Tests** table, the main effect for the within-subjects variable of **TRIAL** is presented first, and is significant, based on all four multivariate tests of significance (Pillai's, Wilks', Hotelling's, and Roy's), $F(3,14)$ = 59.69, $p < .001$. From the cell means presented in the **Estimated Marginal Means** table of **TRIAL**, the results indicate that the subjects made progressively lower number of errors across the four trials, averaged across the two **DRUG** groups and the two **SEX** groups (Trial 1 = 24.55, Trial 2 = 21.10, Trial 3 = 17.55, and Trial 4 = 13.15).

The **Tests of Within-Subjects Contrasts** present the contrasts between the number of errors made across the four trials. The first contrast, LEVEL 1 vs. LEVEL 2, compares the number of errors made in Trial 1 (M = 24.55) with those made in Trial 2 (M = 21.10), and is statistically significant, $F(1,16)$ = 29.66, $p < .001$. This indicates that the subjects made significantly less errors in Trial 2 relative to Trial 1. The second contrast, LEVEL 2 vs. LEVEL 3, compares the number of errors made in Trial 2 (M = 21.10) with the number of errors made in Trial 3 (M = 17.55), and is statistically significant, $F(1,16)$ = 90.02, $p < .001$. This indicates that the subjects made significantly less errors in Trial 3 than in Trial 2. The third contrast, LEVEL 3 vs. LEVEL 4, compares the number of errors made in Trial 3 (M = 17.55) with the number of errors made in Trial 4 (M = 13.15), and is statistically significant, $F(1,16) = 63.74$, $p < .001$. This indicates that the subjects made significantly less errors in Trial 4 than in Trial 3.

Whereas the **Tests of Within-Subjects Contrasts** test for differences between *adjacent levels* within the within-subjects variable of **TRIAL**, the **Pairwise Comparisons** (with Bonferroni correction) test for differences between all four levels. The comparison results presented in the **Pairwise Comparisons** table (under **Estimated Marginal Means)** indicate that (1) the number of errors made in Trial 1 is significantly different (higher) from Trial 2, Trial 3, and Trial 4 ($p < .001$), (2) the number of errors made in Trial 2 is significantly different (higher) from Trial 3 and Trial 4 ($p < .001$), and (3) the number of errors made in Trial 3 is significantly different (higher) from Trial 4 ($p < .001$).

9.5.5.1.2   TRIAL*DRUG

For the **TRIAL*DRUG** interaction, all four multivariate tests of significance (Pillai's, Hotelling's, Wilks', and Roy's) indicate that this interaction is statistically significant, $F(3,14) = 8.26$, $p < .01$, suggesting that the number of errors made across the four trials is dependent upon the presence or absence of the drug. To aid interpretation of the interaction effect, it would be useful to examine the graph presented in Figure 9.4. Note that the graph was plotted from the mean number of errors presented in the **Estimated Marginal Means** table of **DRUG*TRIAL** interaction.

Figure 9.4 shows that in both drug conditions (present/absent), the number of errors made decreased from Trial 1 to Trial 4. However, the rate of decrease is different, with subjects showing a greater rate of decrease in the drug-absent condition across the four trials.

The **Tests of Within-Subjects Contrasts** present the contrasts for the number of errors made across the four trials for the two **DRUG** groups. The first contrast, LEVEL 1 vs. LEVEL 2, compares the number of errors made in Trial 1 with the number of errors made in Trial 2 and is not significant, $F(1,16) = 2.75$, $p > .05$. In conjunction with Figure 9.4, the results indicate that the decrease in the number of errors made between Trial 1 and Trial 2 is similar for the two **DRUG** groups.

	Mean Difference (Trial 1 vs. Trial 2)
Drug present	−2.40 (21.70 − 19.30)
Drug absent	−4.50 (27.40 − 22.90)

The second contrast, LEVEL 2 vs. LEVEL 3, compares the number of errors made in Trial 2 with the number of errors made in Trial 3 and is significant, $F(1,16) = 15.02$, $p < .01$. In conjunction with Figure 9.4, the results indicate that the decrease in the number of errors made between Trial 2 and Trial 3 is different for the two **DRUG** groups; the decrease is greater for the Drug-Absent group.

	Mean Difference (Trial 2 vs. Trial 3)
Drug present	2.10 (19.30 – 17.20)
Drug absent	5.00 (22.90 – 17.90)

The third contrast, LEVEL 3 vs. LEVEL 4, compares the number of errors made in Trial 3 with the number of errors made in Trial 4, and is not significant, $F(1,16) = 0.82$, $p > .05$. In conjunction with Figure 9.4, the results indicate that the decrease in the number of errors made between Trial 3 and Trial 4 is similar for the two DRUG groups.

	Mean Difference (Trial 3 vs. Trial 4)
Drug present	3.90 (17.20 – 13.30)
Drug absent	4.90 (17.90 – 13.00)

9.5.5.1.3   TRIAL*SEX

For the **TRIAL*SEX** interaction, all four multivariate tests of significance (Pillai's, Hotelling's, Wilks', and Roy's) indicate that this interaction is statistically significant, $F(3,14) = 4.45$, $p < .05$, suggesting that the number of errors made across the four trials is dependent upon subjects' gender. To aid interpretation of the interaction effect, it would be useful to examine the graph presented as Figure 9.5. Please note that the graph was plotted from the mean number of errors presented in the **Estimated Marginal Means** table of **SEX*TRIAL** interaction.

Figure 9.5 shows that for both male and female subjects, the number of errors made decreased from Trial 1 to Trial 4. However, the rate of decrease is different, with males showing a greater rate of decrease from Trial 1 to Trial 2.

The Tests of Within-Subjects Contrasts present the contrasts for the number of errors made across the four trials for the male and female groups. The first contrast, LEVEL 1 vs. LEVEL 2, is significant, $F(1,16) = 10.47$, $p < .001$. In conjunction with Figure 9.5, the results indicate that the decrease in the number of errors made between Trial 1 and Trial 2 is different for the male and female groups; the decrease is greater for the male group.

	Mean Difference (Trial 1 vs. Trial 2)
Male	5.50 (20.80 – 15.30)
Female	1.40 (28.30 – 26.90)

The second contrast, LEVEL 2 vs. LEVEL 3, is not significant, $F(1,16) = 0.02$, $p > .05$. In conjunction with Figure 9.5, the results indicate that the decrease in the number of errors made between Trial 2 and Trial 3 is similar for the male and female groups.

	Mean Difference (Trial 2 vs. Trial 3)
Male	3.50 (15.30 – 11.80)
Female	3.60 (26.90 – 23.30)

The third contrast, LEVEL 3 vs. LEVEL 4, compares the number of errors made in Trial 3 with the number of errors made in Trial 4, and is not significant, $F(1,16) = 2.11$, $p > .05$. In conjunction with Figure 9.5, the results indicate that the decrease in the number of errors made between Trial 3 and Trial 4 is similar for the male and female groups.

	Mean Difference (Trial 3 vs. Trial 4)
Male	3.60 (11.80 – 8.20)
Female	5.20 (23.30 – 18.10)

9.5.5.1.4 TRIAL*DRUG*SEX

For the three-way **TRIAL*DRUG*SEX** interaction (presented as **TRIAL*GROUP** in the SPSS output), all four multivariate tests indicate that this interaction is statistically significant ($p < .01$), suggesting that the number of errors made across the four trials is dependent on the **DRUG*SEX** interaction. To aid interpretation of the interaction effect, it would be useful to examine the graph presented as Figure 9.6. Please note that the graph was plotted from the mean number of errors presented in the **Estimated Marginal Means** table of **GROUP*TRIAL** interaction.

Figure 9.6 shows that female subjects made more errors under the drug-absent condition than under the drug-present condition across the first three trials. However, for the last trial, the effect is opposite with more errors made under the drug-present condition than under the drug-absent condition. For male subjects, the response pattern was similar to that of the female subjects across the first two trials (i.e., more errors were made under the drug-absent condition than under the drug-present condition). However, for the last two trials, there was very little difference in the number of errors made under these two conditions.

For this three-way interaction, the **Tests of Within-Subjects Contrasts** showed that the first contrast, LEVEL 1 vs. LEVEL 2, and the second contrast, LEVEL 2 vs. LEVEL 3, are significant, $F(3,16) = 4.76$, $p < .05$ and $F(3,16) = 5.49$, $p < .01$, respectively. In conjunction with Figure 9.6, the results indicate that the mean differences in the number of errors made between Trial 1 and Trial 2 and between Trial 2 and Trial 3 are different for the four groups (**drug-present male**, **drug-present female**, **drug-absent male**, and **drug-absent female**). For the male subjects, the decrease in the number of errors made from Trial 1 to Trial 2 is greater under the drug-absent condition (M = 7.20) than under the drug-present condition (M = 3.80). For the female subjects, the decrease in the number of errors made from Trial 1 to Trial 2 is similar under both the drug-absent (M = 1.80) and the drug-present conditions (M

= 1.00). The results show that between Trial 2 and Trial 3, the decrease in the number of errors made is greater under the drug-absent condition than under the drug-present condition for both male and females. However, the decrease in errors between the drug-absent and drug-present conditions is greater for males than for females.

	Mean Difference (Trial 1 vs. Trial 2)
Drug-present male	3.80 (17.40 – 13.60)
Drug-absent male	7.20 (24.20 – 17.00)
Drug-present female	1.00 (26.00 – 25.00)
Drug-absent female	1.80 (30.60 – 28.80)

	Mean Difference (Trial 2 vs. Trial 3)
Drug-present male	1.60 (13.60 – 12.00)
Drug-absent male	5.40 (17.00 – 11.60)
Drug-present female	2.60 (25.00 – 22.40)
Drug-absent female	4.60 (28.80 – 24.20)

The results indicate that the last contrast, LEVEL 3 vs. LEVEL 4, is not significant, $F(3,16) = 2.83$, $p > .05$. In conjunction with Figure 9.6, the results indicate that the decrease in the number of errors made between Trial 3 and Trial 4 is similar for the four groups of drug-present male, drug-present female, drug-absent male, and drug-absent female.

	Mean Difference (Trial 3 vs. Trial 4)
Drug-present male	4.40 (12.00 – 7.60)
Drug-absent male	2.80 (11.60 – 8.80)
Drug-present female	3.40 (22.40 – 19.00)
Drug-absent female	7.00 (24.20 – 17.20)

9.5.5.2   Between-Groups Effects

The final analysis presented is the **Tests for Between-Subjects Effects**. This is equivalent to a two-way ANOVA. The results indicate that only subjects' **SEX** exerted a significant effect on the number of errors made, $F(1,16) = 24.45$, $p < .001$. From the **Estimated Marginal Means** table of **SEX**, it can be seen that, on average (i.e., collapsing across the **DRUG** and **TRIAL** conditions), females made more errors than males (females: M = 24.15, males: M = 14.02). Neither the **DRUG** variable, $F(1,16) = 1.40$, $p > .05$, nor the **DRUG*SEX** interaction, $F(1,16) = 0.03$, $p < .05$, had a significant effect on the number of errors made. In conjunction with Figure 9.3, the results indicate that, regardless of the presence or absence of the drug, females made more errors than males, averaged across the four trials.

10

Correlation

10.1 Aim

Correlation is primarily concerned with finding out whether a relationship exists and with determining its magnitude and direction. When two variables vary together, such as loneliness and depression, they are said to be *correlated*. Accordingly, correlational studies are attempts to find the extent to which two or more variables are related. Typically, in a correlational study, no variables are manipulated as in an experiment — the researcher measures naturally occurring events, behaviors, or personality characteristics and then determines if the measured scores covary. The simplest correlational study involves obtaining a pair of observations or measures on two different variables from a number of individuals. The paired measures are then statistically analyzed to determine if any relationship exists between them. For example, behavioral scientists have explored the relationship between variables such as anxiety level and self-esteem, attendance at classes in school and course grades, university performance and career success, and body weight and self-esteem.

To quantitatively express the extent to which two variables are related, it is necessary to calculate a *correlation coefficient*. There are many types of correlation coefficients, and the decision of which one to employ with a specific set of data depends on the following factors:

- The level of measurement on which each variable is measured
- The nature of the underlying distribution (continuous or discrete)
- The characteristics of the distribution of the scores (linear or non-linear)

This chapter presents two correlation coefficients: the *Pearson product moment correlation coefficient* (r), employed with interval or ratio scaled variables, and the *Spearman rank order correlation coefficient* (r_{rho}), employed with ordered or ranked data. It is important to note that, regardless of which correlational technique the researcher uses, they all have the following characteristics in common:

1. Two sets of measurements are obtained on the same individuals or on pairs of individuals who are matched on some basis.
2. The values of the correlation coefficients vary between +1.00 and −1.00. Both of these extremes represent perfect relationships between the variables, and 0.00 represents the absence of a relationship.
3. A *positive relationship* means that individuals obtaining high scores on one variable tend to obtain high scores on a second variable. The converse is also true, i.e., individuals scoring low on one variable tend to score low on a second variable.
4. A *negative relationship* means that individuals scoring low on one variable tend to score high on a second variable. Conversely, individuals scoring high on one variable tend to score low on a second variable.

10.2 Assumptions

- For each subject in the study, there must be *related pairs of scores,* i.e., if a subject has a score on variable X, then the same subject must also have a score on variable Y.
- The relationship between the two variables must be *linear,* i.e., the relationship can be most accurately represented by a straight line.
- The variables should be measured at least at the *ordinal level.*
- The variability of scores on the Y variable should remain constant at all values of the X variable. This assumption is called *homoscedasticity.*

10.3 Example 1 — Pearson Product Moment Correlation Coefficient

Assume that a researcher wishes to determine whether there is a relationship between grade point average (GPA) and the scores on a reading-comprehension test of 15 first-year students. The researcher recorded the pair of scores given in the following, together with their rankings:

Student	Read	Read_rank	GPA	GPA_rank
s1	38	13	2.1	13
s2	54	3	2.9	6
s3	43	10	3.0	5
s4	45	8	2.3	12
s5	50	4	2.6	7.5
s6	61	1	3.7	1
s7	57	2	3.2	4
s8	25	15	1.3	15
s9	36	14	1.8	14
s10	39	11.5	2.5	9.5
s11	48	5.5	3.4	2
s12	46	7	2.6	7.5
s13	44	9	2.4	11
s14	39	11.5	2.5	9.5
s15	48	5.5	3.3	3

10.3.1   Data Entry Format

The data set has been saved under the name: **CORR.SAV**.

Variables	Column(s)	Code
READ	1	Reading score
READ_RANK	2	Ranking
GPA	3	Grade point average
GPA_RANK	4	Ranking

Step 1: To test the assumptions of linearity and homoscedasticity, a scatter plot will be obtained.

10.3.2   Scatterplot (Windows Method)

1. From the menu bar, click **Graphs**, and then **Scatter/Dot...** The following **Scatter/Dot** window will open. Click (highlight) the

 icon.

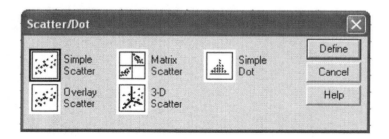

2. Click **Define** to open the **Simple Scatterplot** window.

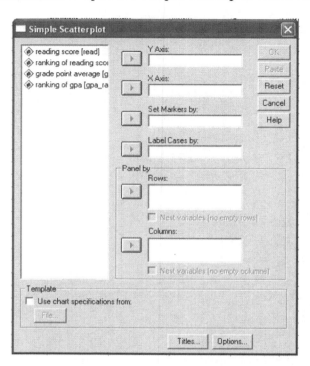

3. Transfer the **READ** variable to the **Y Axis** field by clicking (high-lighting) the variable and then clicking [▶]. Transfer the **GPA variable** to the **X Axis** field by clicking (highlighting) the variable and then clicking [▶].

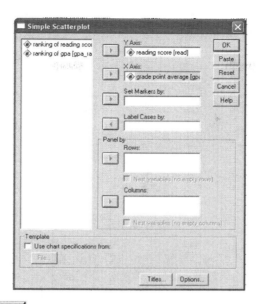

4. Click **Options...** to open the **Options** window. Under **Missing Values**, ensure that the **Exclude cases listwise** field is checked. By default, for scatterplot, SPSS employs the **listwise** method for handling missing data. (For a discussion on the differences in the pairwise and the listwise methods of handling missing data, please see Subsection 10.3.6.)

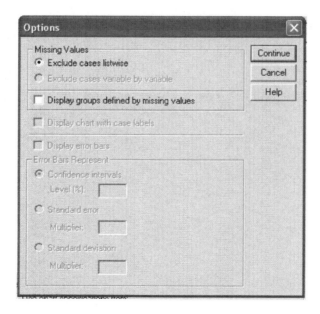

5. Click [**Continue**] to return to the **Simple Scatterplot** window.

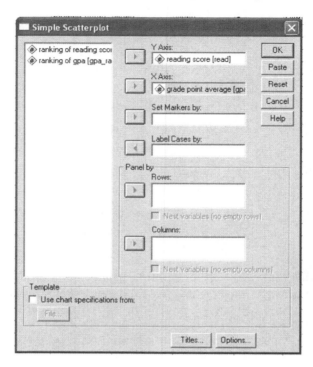

6. When the **Simple scatterplot** window opens, click [**OK**] to complete the analysis. See Figure 10.1 for the results.

10.3.3   Scatterplot (SPSS Syntax Method)

```
GRAPH
/SCATTERPLOT(BIVAR)=READ WITH GPA
/MISSING=LISTWISE.
```

10.3.4   Scatterplot

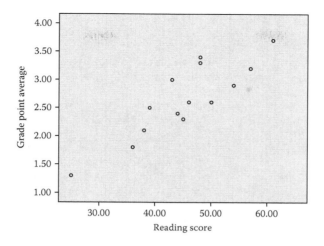

FIGURE 10.1

10.3.5   Interpretation

As can be seen from Figure 10.1, there is a linear relationship between the variables of reading score and GPA, such that as the reading score increases, so does GPA. The figure also shows that the homoscedasticity assumption is met, because the variability of the GPA score remains relatively constant from one READ score to the next.

Step 2: Pearson product moment correlation.

10.3.6   Pearson Product Moment Correlation (Windows Method)

1. From the menu bar, click **Analyze**, then **Correlate,** and then **Bivariate....** The following **Bivariate Correlations** window will open.

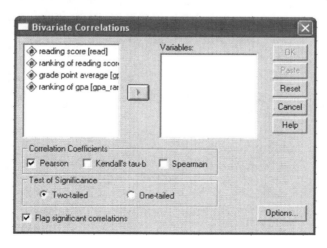

2. Transfer the **READ** and **GPA** variables to the **Variables** field by clicking (highlighting) them and then clicking 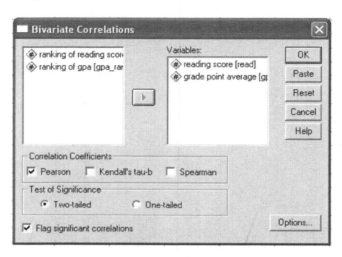. By default, SPSS will employ the **Pearson correlation analysis**, and a **two-tailed test of significance** (both fields are checked).

3. Click 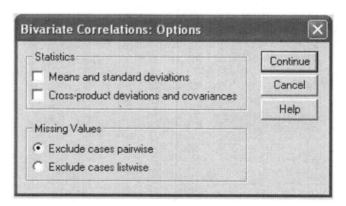 to open the **Bivariate Correlation: Options** window.

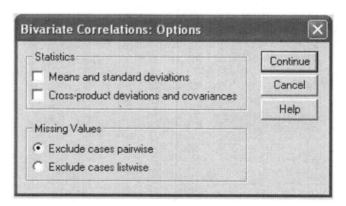

By default, SPSS employs the **Exclude cases pairwise** method for handling missing data (this field is checked). This method treats the calculation of each pair of variables as a separate problem, using all cases with complete data for the pair. With this method, only cases with missing data on a specific pair of variables will be excluded from the calculation. As such, the correlation for different pairs of variables may be based on different number of cases. As an option, the **Exclude cases listwise** method can be used to include only those cases with complete data. With this method, any case with missing data on any pair of variables will be excluded from the calculation. As such, the sample size for the correlation between any pair of variables can be reduced further with the listwise method.

4. Click [Continue] to return to the **Bivariate Correlations** window.

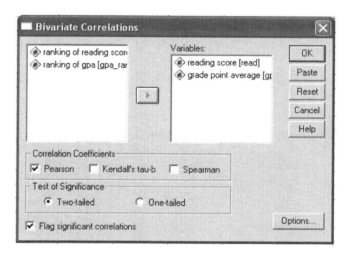

5. When the Bivariate Correlations window opens, click [OK] to complete the analysis. See Table 10.1 for the results.

10.3.7 Pearson Product Moment Correlation (SPSS Syntax Method)

CORRELATIONS READ WITH GPA
/MISSING=PAIRWISE.

10.3.8 SPSS Output

TABLE 10.1

Pearson Product Moment Correlation

Correlations		
		grade point average
Reading score	Pearson Correlation	.867
	Sig. (2-tailed)	.000
	N	15

10.3.9   Results and Interpretation

The correlation between reading score and GPA is positive and statistically significant ($r = 0.867$, $p < .001$). This means that as the students' reading scores increase, so do their GPA scores. Please note that this interpretation in no way implies *causality* — that increases in reading scores caused increases in GPA scores. The significant relationship merely indicates that the two variables *covary*.

10.4   Example 2 — Spearman Rank Order Correlation Coefficient

For this example, the same data set (**CORR.SAV**) will be used. However, the rank order of the two variables (READ_RANK, GPA_RANK) will be used instead of their actual values as recorded. Thus, the computation for this coefficient is not sensitive to asymmetrical distributions or to the presence of outliers.

10.4.1   Windows Method

1. From the menu bar, click **Analyze**, then **Correlate**, and then **Bivariate....** The following **Bivariate Correlations** window will open.

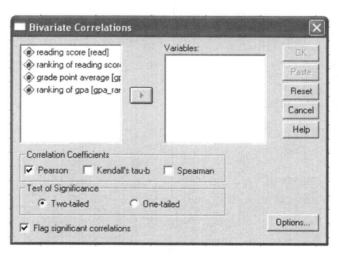

2. Transfer the **READ_RANK** and **GPA_RANK** variables to the **Variables** field by clicking (highlighting) them and then clicking .

By default, SPSS will employ the **Pearson correlation analysis** (this field is checked). Uncheck the **Pearson** field and check the **Spearman** field.

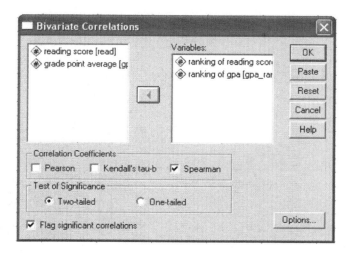

3. Click 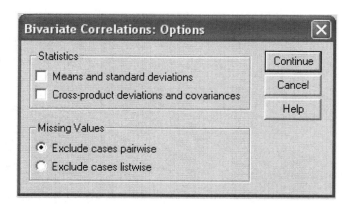 to open the **Bivariate Correlation: Options** window. Ensure that the **Exclude cases pairwise** field is checked.

4. Click [Continue] to return to the **Bivariate Correlations** window.

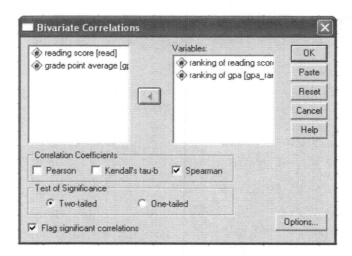

5. When the Bivariate Correlations window opens, click [OK] to complete the analysis. See Table 10.2 for the results.

10.4.2 SPSS Syntax Method

NONPAR CORR READ_RANK WITH GPA_RANK.

10.4.3 SPSS Output

TABLE 10.2

Spearman Rank Order Correlation

Correlations			
			ranking of gpa
Spearman's rho	Ranking of reading scores	Correlation Coefficient	.826
		Sig. (2-tailed)	.000
		N	15

10.4.4 Results and Interpretation

The obtained Spearman rank-order coefficient (rho = 0.826, $p < .001$) is highly similar in magnitude and direction to that in the Pearson correlation table (Table 10.1). Thus, similar to the Pearson coefficient, the Spearman coefficient indicates that as the students' reading scores increase, so do their GPA scores.

11

Linear Regression

11.1 Aim

Regression and correlation are closely related. Both techniques involve the relationship between two variables, and they both utilize the same set of paired scores taken from the same subjects. However, whereas correlation is concerned with the magnitude and direction of the relationship, regression focuses on using the relationship for *prediction*. In terms of prediction, if two variables were correlated perfectly, then knowing the value of one score permits a perfect prediction of the score on the second variable. Generally, whenever two variables are significantly correlated, the researcher may use the score on one variable to predict the score on the second.

There are many reasons why researchers want to predict one variable from another. For example, knowing a person's I.Q., what can we say about this person's prospects of successfully completing a university course? Knowing a person's prior voting record, can we make any informed guesses concerning his vote in the coming election? Knowing his mathematics aptitude score, can we estimate the quality of his performance in a course in statistics? These questions involve predictions from one variable to another, and psychologists, educators, biologists, sociologists, and economists are constantly being called upon to perform this function.

11.2 Assumptions

- For each subject in the study, there must be *related pairs of scores*. That is, if a subject has a score on variable X, then the same subject must also have a score on variable Y.
- The relationship between the two variables must be *linear*, i.e., the relationship can be most accurately represented by a straight line.

- The variables should be measured at least at the *ordinal* level.
- The variability of scores on the Y variable should remain constant at all values of the X variable. This assumption is called *homoscedasticity.*

11.3 Example — Linear Regression

This example employs the data set **CORR.SAV** from the previous chapter (see Section 10.3 in Chapter 10). In this example, we wish to (1) find the prediction equation that allows us to best predict the students' grade point average (GPA) scores from their reading scores (**READ**), (2) determine the strength of this relationship, and (3) test the null hypothesis that **READ** and **GPA** scores are unrelated.

11.3.1 Windows Method

1. From the menu bar, click **Analyze**, then **Regression**, and then **Linear....** The following **Linear Regression** window will open.

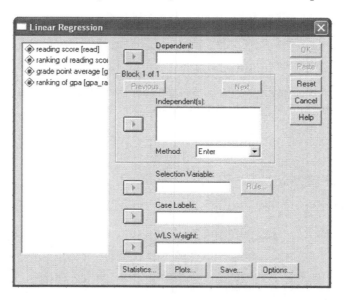

2. Click (highlight) the **GPA** variable and then click ▶ to transfer this variable to the **Dependent** field. Next, click (highlight) the

READ variable and then click 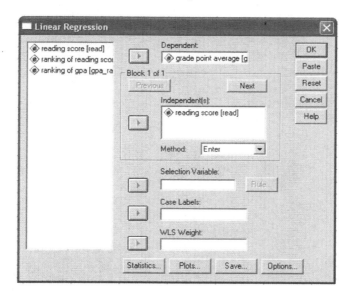 to transfer this variable to the **Independent(s)** field. In the **Method** field, select **ENTER** from the drop-down list as the method of entry for the independent (predictor) variable into the prediction equation.

3. Click `Statistics...` to open the **Linear Regression: Statistics** window. Check the fields to obtain the statistics required. For this example, check the fields for **Estimates, Confidence intervals,** and **Model fit**.

Click `Continue` when finished.

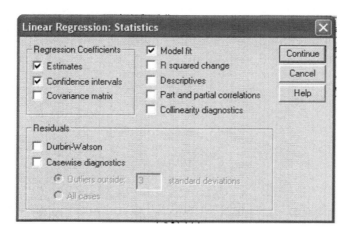

4. When the **Linear Regression** window opens, click [Options...] to open the **Linear Regression: Options** window. Ensure that both the **Use probability of F** and the **Include constant in equation** fields are checked. Click [Continue] to return to the **Linear Regression** window.

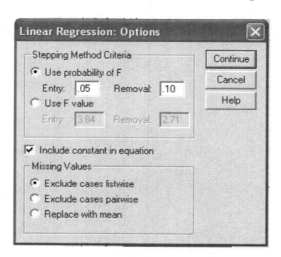

5. When the **Linear Regression** window opens, click [OK] to complete the analysis. See Table 11.1 for the results.

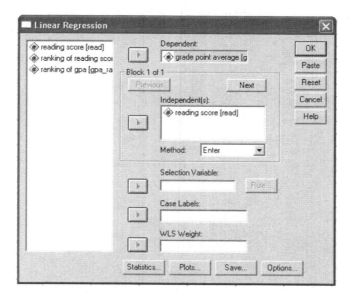

11.3.2 SPSS Syntax Method

REGRESSION VARIABLES=(COLLECT)
/STATISTICS=DEFAULTS CI
/DEPENDENT=GPA
/METHOD=ENTER READ.

11.3.3 SPSS Output

TABLE 11.1

Linear Regression Analysis Output

Regression

Variables Entered/Removed[b]

Model	Variables Entered	Variables Removed	Method
1	Reading score[a]		Enter

[a]All requested variables entered.
[b]Dependent variable: grade point average.

Model Summary

Model	R	R Square	Adjusted R Square	Std. Error of the Estimate
1	.867[a]	.752	.733	.32848

[a]Predictors: (constant), reading score.

ANOVA[b]

Model		Sum of Squares	df	Mean Square	F	Sig.
1	Regression	4.253	1	4.253	39.418	.000[a]
	Residual	1.403	13	.108		
	Total	5.656	14			

[a]Predictors: (constant), reading score.
[b]Dependent variable: grade point average.

Coefficients[a]

Model		Unstandardized Coefficients B	Unstandardized Coefficients Std. Error	Standardized Coefficients Beta	t	Sig.	95% Confidence Interval for B Lower Bound	95% Confidence Interval for B Upper Bound
1	(Constant)	−.111	.448		−.248	.808	−1.075	.853
	reading score	.061	.010	.867	6.278	.000	.040	.082

[a.]Dependent variable: grade point average.

11.3.4   Results and Interpretation

11.3.4.1   *Prediction Equation*

The prediction equation is: **Y' = A+BX**, where Y' = the predicted dependent variable, A = Constant, B = unstandardized regression coefficient, and X = value of the predictor variable.

The relevant information for constructing a least-squares regression (prediction) equation is presented in the **Coefficients** table (see Table 11.1). To predict the students' GPA scores from their reading scores (**READ**), use the values presented in the **Unstandardized Coefficients** column. Using the **Constant** and **B** (unstandardized coefficient) values, the prediction equation would be:

$$\text{Predicted GPA} = -0.111 + (0.061 \times \text{READ})$$

Thus, for a student who has a reading score of 56, his/her predicted GPA score will be:

$$\text{Predicted GPA} = -0.111 + (0.061 \times 56)$$

$$= 3.31$$

However, in the **Model Summary** table, the **Standard Error of the Estimate** is 0.32848. This means that at the 95% confidence interval, the predicted GPA score of 3.31 lies between the scores of **2.66** ($3.31 - (1.96 \times 0.32848)$) and **3.95** ($3.31 + (1.96 \times 0.32848)$).

11.3.4.2   *Evaluating the Strength of the Prediction Equation*

A measure of the strength of the computed equation is **R-square**, sometimes called the **coefficient of determination**. R-square is simply the square of the multiple correlation coefficient listed under **R** in the **Model Summary** table, and represents the proportion of variance accounted for in the dependent variable (GPA) by the predictor variable (READ). In a simple regression such as this, where there is only one predictor variable, the multiple R is equivalent to the simple R (Pearson product–moment correlation). For this example, the multiple correlation coefficient is 0.867, and the R-square is 0.752. Thus, for this sample, the predictor variable of READ has explained 75.2% of the variance in the dependent variable of GPA.

The **ANOVA** table presents results from the test of the null hypothesis that R-square is zero. An R-square of zero indicates no linear relationship between the predictor and dependent variable. The **ANOVA** table shows that the computed F statistic is 39.42, with an observed significance level of less than 0.001. Thus, the hypothesis that there is no linear relationship between the predictor and dependent variable is rejected.

11.3.4.3   Identifying Independent Relationship

The **Coefficients** table presents the standardized **Beta** coefficient between the predictor variable READ and the dependent variable GPA. The Beta coefficient is shown to be positive and statistically significant at the 0.001 level. Thus, the higher the students' reading scores, the higher their GPA scores, Beta = 0.87, t = 6.28, $p < .001$. Note that the standardized Beta coefficient of 0.87 is identical to the multiple R coefficient. This is because there is only one predictor variable.

12

Factor Analysis

12.1 Aim

The major aim of factor analysis is the orderly simplification of a large number of intercorrelated measures to a few representative constructs or factors. Suppose that a researcher wants to identify the major dimensions underlying a number of personality tests. He begins by administering the personality tests to a large sample of people (N = 1000), with each test supposedly measuring a specific facet of a person's personality (e.g., ethnocentrism, authoritarianism, and locus of control). Assume that there are 30 such tests, each consisting of ten test items. What the researcher will end up with is a mass of numbers that will say very little about the dimensions underlying these personality tests. On average, some of the scores will be high, some will be low, and some intermediate, but interpretation of these scores will be extremely difficult if not impossible. This is where factor analysis comes in. It allows the researcher to "reduce" this mass of numbers to a few representative factors, which can then be used for subsequent analysis.

Factor analysis is based on the assumption that all variables are correlated to some degree. Therefore, those variables that share similar underlying dimensions should be highly correlated, and those variables that measure dissimilar dimensions should yield low correlations. Using the earlier example, if the researcher intercorrelates the scores obtained from the 30 personality tests, then those tests that measure the same underlying personality dimension should yield high correlation coefficients, whereas those tests that measure different personality dimensions should yield low correlation coefficients. These high/low correlation coefficients will become apparent in the correlation matrix because they form clusters indicating which variables "hang" together. For example, measures of ethnocentrism, authoritarianism, and aggression may be highly intercorrelated, indicating that they form an identifiable personality dimension. The primary function of factor analysis is to identify these clusters of high intercorrelations as independent factors.

There are three basic steps to factor analysis:

1. Computation of the correlation matrix for all variables.
2. Extraction of initial factors.
3. Rotation of the extracted factors to a terminal solution.

12.1.1   Computation of the Correlation Matrix

As factor analysis is based on correlations between measured variables, a correlation matrix containing the intercorrelation coefficients for the variables must be computed. The variables should be measured at least at the ordinal level, although two-category nominal variables (coded 1-2) can be used. If all variables are nominal variables, then specialized forms of factor analysis, such as Boolean factor analysis (BMPD, 1992), are more appropriate.

12.1.2   Extraction of Initial Factors

At this phase, the number of common factors needed to adequately describe the data is determined. To do this, the researcher must decide on (1) the method of extraction, and (2) the number of factors selected to represent the underlying structure of the data.

12.1.2.1   Method of Extraction

There are two basic methods for obtaining factor solutions. They are **Principal Components** analysis and common **Factor Analysis** (Note: SPSS provides six methods of extraction under the common factor analysis model; these are: Principal-axis factoring, unweighted least-squares, generalized least-squares, maximum-likelihood, alpha factoring, and image factoring.) The choice between these two basic methods of factor extraction lies with the objective of the researcher. If the purpose is no more than to "reduce data" to obtain the minimum number of factors needed to represent the original set of data, then **Principal Components** analysis is appropriate. The researcher works from the premise that the factors extracted need not have any theoretical validity. Conversely, when the primary objective is to identify theoretically meaningful underlying dimensions, the common **Factor Analysis** method is the appropriate model. Given the more restrictive assumptions underlying common factor analysis, the principal components method has attracted more widespread use.

12.1.2.2   Determining the Number of Factors to Be Extracted

There are two conventional criteria for determining the number of initial unrotated factors to be extracted. These are the **Eigenvalues** criterion and the **Scree test** criterion.

12.1.2.2.1 Eigenvalues

Only factors with eigenvalues of 1 or greater are considered to be significant; all factors with eigenvalues less than 1 are disregarded. An eigenvalue is a *ratio* between the common (shared) variance and the specific (unique) variance explained by a specific factor extracted. The rationale for using the eigenvalue criterion is that the amount of common variance explained by an extracted factor should be at least equal to the variance explained by a single variable (unique variance) if that factor is to be retained for interpretation. An eigenvalue greater than 1 indicates that more common variance than unique variance is explained by that factor.

12.1.2.2.2 Scree Test

This test is used to identify the optimum number of factors that can be extracted before the amount of unique variance begins to dominate the common variance structure (Hair, Anderson, Tatham, & Black, 1995). The scree test is derived by plotting the eigenvalues (on the Y axis) against the number of factors in their order of extraction (on the X axis). The initial factors extracted are large factors (with high eigenvalues), followed by smaller factors. Graphically, the plot will show a steep slope between the large factors and the gradual trailing off of the rest of the factors. The point at which the curve first begins to straighten out is considered to indicate the maximum number of factors to extract. That is, those factors above this point of inflection are deemed meaningful, and those below are not. As a general rule, the scree test results in at least one and sometimes two or three more factors being considered significant than does the eigenvalue criterion (Cattel, 1966).

12.1.3 Rotation of Extracted Factors

Factors produced in the initial extraction phase are often difficult to interpret. This is because the procedure in this phase ignores the possibility that variables identified to load on or represent factors may already have high loadings (correlations) with previous factors extracted. This may result in significant cross-loadings in which many factors are correlated with many variables. This makes interpretation of each factor difficult, because different factors are represented by the same variables. The rotation phase serves to "sharpen" the factors by identifying those variables that load on one factor and not on another. The ultimate effect of the rotation phase is to achieve a simpler, theoretically more meaningful factor pattern.

12.1.4 Rotation Methods

There are two main classes of factor rotation method: **Orthogonal** and **Oblique**. Orthogonal rotation assumes that the factors are independent, and the rotation process maintains the reference axes of the factors at 90°. Oblique rotation allows for correlated factors instead of maintaining independence between the rotated factors. The oblique rotation process does not require that the reference axes be maintained at 90°. Of the two rotation methods, oblique rotation is more flexible because the factor axes need not be orthogonal. Moreover, at the theoretical level, it is more realistic to assume that influences in nature are correlated. By allowing for correlated factors, oblique rotation often represents the clustering of variables more accurately.

There are three major methods of orthogonal rotation: **varimax**, **quartimax**, and **equimax**. Of the three approaches, varimax has achieved the most widespread use as it seems to give the clearest separation of factors. It does this by producing the maximum possible simplification of the columns (factors) within the factor matrix. In contrast, both quartimax and equimax approaches have not proven very successful in producing simpler structures, and have not gained widespread acceptance. Whereas the orthogonal approach to rotation has several choices provided by SPSS, the oblique approach is limited to one method: **oblimin**.

12.1.5   Orthogonal vs. Oblique Rotation

In choosing between orthogonal and oblique rotation, there is no compelling analytical reason to favor one method over the other. Indeed, there are no hard and fast rules to guide the researcher in selecting a particular orthogonal or oblique rotational technique. However, convention suggests that the following guidelines may help in the selection process. If the goal of the research is no more than to "reduce the data" to more manageable proportions, regardless of how meaningful the resulting factors may be, and if there is reason to assume that the factors are uncorrelated, then orthogonal rotation should be used. Conversely, if the goal of the research is to discover theoretically meaningful factors, and if there are theoretical reasons to assume that the factors will be correlated, then oblique rotation is appropriate.

Sometimes the researcher may not know whether or not the extracted factors might be correlated. In such a case, the researcher should try an oblique solution first. This suggestion is based on the assumption that, realistically, very few variables in a particular research project will be uncorrelated. If the correlations between the factors turn out to be very low (e.g., < 0.20), the researcher could redo the analysis with an orthogonal rotation method.

12.1.6   Number of Factor Analysis Runs

It should be noted that when factor analysis is used for research (either for the purpose of data reduction or to identify theoretically meaningful dimensions), a minimum of two runs will normally be required. In the first run, the researcher allows factor analysis to extract factors for rotation. All factors with eigenvalues of 1 or greater will be subjected to varimax rotation by default within SPSS. However, even after rotation, not all extracted rotated factors will be meaningful. For example, some small factors may be represented by very few items, and there may still be significant cross-loading of items across several factors. At this stage, the researcher must decide which factors are substantively meaningful (either theoretically or intuitively), and retain only these for further rotation. It is not uncommon for a data set to be subjected to a series of factor analysis and rotation before the obtained factors can be considered "clean" and interpretable.

12.1.7 Interpreting Factors

In interpreting factors, the size of the factor loadings (correlation coefficients between the variables and the factors they represent) will help in the interpretation. As a general rule, variables with large loadings indicate that they are representative of the factor, while small loadings suggest that they are not. In deciding what is large or small, a rule of thumb suggests that factor loadings greater than ± 0.33 are considered to meet the minimal level of practical significance. The reason for using the ± 0.33 criterion is that if the value is squared, the squared value represents the amount of the variable's total variance accounted for by the factor. Therefore, a factor loading of 0.33 denotes that approximately 10% of the variable's total variance is accounted for by the factor. The grouping of variables with high factor loadings should suggest what the underlying dimension is for that factor.

12.2 Checklist of Requirements

- Variables for factor analysis should be measured at least at the ordinal level.
- If the researcher has some prior knowledge about the factor structure, then several variables (five or more) should be included to represent each proposed factor.
- The sample size should be 100 or larger. A basic rule of thumb is to have at least five times as many cases as variables entered into the factor analysis. A more acceptable range would be a ten-to-one ratio.

12.3 Assumptions

The assumptions underlying factor analysis can be classified as **statistical** and **conceptual**.

12.3.1 Statistical Assumptions

Statistical assumptions include *normality and linearity* and *sufficient significant correlations in data matrix.*

- **Normality and linearity:** Departures from normality and linearity can diminish the observed correlation between measured variables and thus degrade the factor solution.
- **Sufficient significant correlations in data matrix:** The researcher must ensure that the data matrix has sufficient correlations to justify the application of factor analysis. If visual inspection reveals no substantial number of correlations of 0.33 or greater, then factor analysis is probably inappropriate.

12.3.2 Conceptual Assumptions

Conceptual assumptions include *selection of variables* and *homogeneity.*

- **Selection of variables:** Variables should be selected to reflect the underlying dimensions that are hypothesized to exist in the set of selected variables. This is because factor analysis has no means to determine the appropriateness of the selected variables other than the correlations among the variables.
- **Homogeneity:** The sample must be homogeneous with respect to the underlying factor structure. If the sample consists of two or more distinct groups (e.g., males and females), separate factor analysis should be performed.

12.4 Factor Analysis: Example 1

A study was designed to investigate how the defense of **self defense, provocation**, and **insanity** influence jurors' verdict judgments in trials of battered women who killed their abusive spouses (Ho & Venus, 1995). Nine statements were written to reflect these three defense strategies. Each statement was rated on an 8-point scale, with high scores indicating strong support for that particular defense strategy. A total of 400 subjects provided responses to these nine statements. Factor analysis (with principal components extraction) was employed to investigate whether these nine "defense" statements represent identifiable factors, i.e., defense strategies. The nine statements (together with their SPSS variable name) written to reflect the three defense strategies are listed in the following.

1. Provocation Defense
   - **PROVO:** In killing her husband, the defendant's action reflects a sudden and temporary loss of self-control as a result of the provocative conduct of the deceased.
   - **CAUSED:** The nature of the provocative conduct by the deceased was such that it could have caused an ordinary person with normal powers of self-control to do what the defendant did.
   - **PASSION:** In killing her husband, the defendant acted in the heat of passion as a response to the deceased sudden provocation on that fateful day.

2. Self-Defense Defense
   - **PROTECT:** In killing her husband, the defendant was justified in using whatever force (including lethal force) to protect herself.
   - **SAVE:** The defendant's lethal action was justified in that she acted to save herself from grievous bodily harm.
   - **DEFEND:** In killing her husband, the defendant used such force as was necessary to defend herself.

3. Insanity Defense
   - **MENTAL:** The action of the defendant is the action of a mentally impaired person.
   - **INSANE:** In killing her husband, the defendant was either irrational or insane.
   - **STABLE:** The action of the accused is not typical of the action of a mentally stable person.

12.4.1 Data Entry Format

Note: The survey questionnaire employed in this study was designed as part of a larger study. It contains additional variables apart from the nine defense strategy variables. The data set is named **DOMES.SAV**.

Variables	Column(s)	Code
Sex	1	1 = male, 2 = female
Age	2	In years
Educ	3	1 = primary to 6 = tertiary
Income	4	1 = < \$10,000 per year, 5 = ≥ \$40,000 per year
Provo	5	1 = strongly disagree, 8 = strongly agree
Protect	6	1 = strongly disagree, 8 = strongly agree
Mental	7	1 = strongly disagree, 8 = strongly agree
Caused	8	1 = strongly disagree, 8 = strongly agree
Save	9	1 = strongly disagree, 8 = strongly agree
Insane	10	1 = strongly disagree, 8 = strongly agree
Passion	11	1 = strongly disagree, 8 = strongly agree
Defend	12	1 = strongly disagree, 8 = strongly agree
Stable	13	1 = strongly disagree, 8 = strongly agree
Real	14	1 = totally unbelievable syndrome, 8 = totally believable syndrome
Support	15	1 = no support at all, 8 = a great deal of support
Apply	16	1 = does not apply to the defendant at all, 8 = totally applies to the defendant
Respon	17	1 = not at all responsible, 8 = entirely responsible
Verdict	18	1 = not guilty, 2 = guilty of manslaughter, 3 = guilty of murder
Sentence	19	1 = 0 yr, 6 = life imprisonment

12.4.2   Windows Method

1. From the menu bar, click **Analyze**, then **Data Reduction**, and then **Factor**. The following **Factor Analysis** window will open.

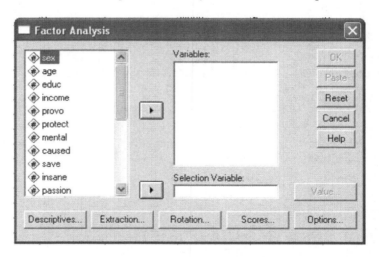

2. Transfer the nine variables of **PROVO, PROTECT, MENTAL, CAUSED, SAVE, INSANE, PASSION, DEFEND,** and **STABLE** to

the **Variables** field by clicking these variables (highlight) and then clicking 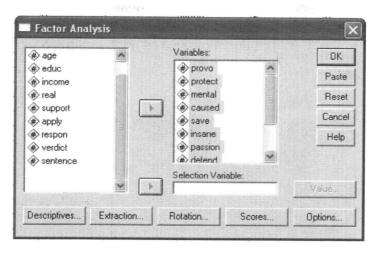 .

3. To (1) obtain a correlation matrix for the nine variables, and (2) to test that the correlation matrix has sufficient correlations to justify the application of factor analysis, click **Descriptives...** . The following **Factor Analysis: Descriptives** window will open. Check the **Coefficients** and **KMO and Bartlett's test of sphericity** fields, and then click **Continue** .

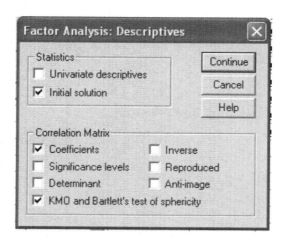

4. When the **Factor Analysis** window opens, click ⌐Extraction...⌐ . This will open the **Factor Analysis: Extraction** window. In the **Method:** drop-down list, choose **Principal components** as the extraction method. In the **Eigenvalues over** field, accept the default value of **1**. Leave the **Number of factors** field blank (i.e., allow principal components analysis to extract as many factors as there are with eigenvalues greater than 1). To obtain a **Scree plot** of the number of factors extracted, check the **Scree plot** field. Click ⌐Continue⌐ .

Factor Analysis: Extraction

Method:	Principal components ▼		Continue
Analyze		**Display**	Cancel
⦿ Correlation matrix		☑ Unrotated factor solution	Help
○ Covariance matrix		☑ Scree plot	

Extract
⦿ Eigenvalues over: 1
○ Number of factors:

Maximum Iterations for Convergence: 25

5. When the **Factor Analysis** window opens, click ⌐Rotation...⌐ . This will open the **Factor Analysis: Rotation** window. To subject the extracted factors to **Varimax** rotation, check the **Varimax** field. Click ⌐Continue⌐ .

Factor Analysis: Rotation

Method
○ None   ○ Quartimax   Continue
⦿ Varimax   ○ Equamax   Cancel
○ Direct Oblimin ○ Promax   Help
Delta: 0   Kappa 4

Display
☑ Rotated solution   ☐ Loading plot(s)

Maximum Iterations for Convergence: 25

6. When the **Factor Analysis** window opens, click Options... . This will open the **Factor Analysis: Options** window. If the data set has **missing values**, the researcher can choose one of the three methods offered to deal with the missing values: (1) **Exclude cases listwise** — any case (subject) with a missing value for any of the variables in the factor analysis will be excluded from the analysis; this method is the most restrictive as the presence of missing values can reduce the sample size substantially, (2) **Exclude cases pairwise** — any variable in the factor analysis that has a missing value will be excluded from the analysis; this method is less restrictive as only variables (with missing values), rather than cases, are excluded, and (3) **Replace with mean** — all missing values are replaced with mean values; this method is less restrictive as all variables in the factor analysis will be included in the analysis. Under **Coefficient Display Format**, check the **Sorted by size** field. This procedure will present the factor loadings (correlation coefficients) in a descending order of magnitude format in the output. Check the **Suppress absolute values less than** field, and then type the coefficient **0.33** in the field next to it. This procedure will suppress the presentation of any factor loadings with values less than 0.33 in the output (an item with a factor loading of 0.33 or higher indicates that approximately 10% or more of the variance in that item is accounted for by its common factor). Click Continue .

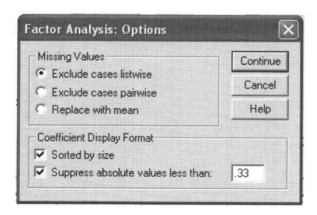

7. When the **Factor Analysis** window opens, click OK to complete the analysis. See Table 12.1 for the results.

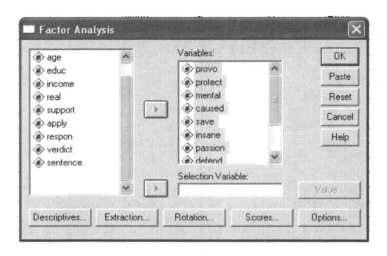

12.4.3   SPSS Syntax Method

FACTOR VARIABLES=PROVO TO STABLE
/FORMAT=SORT BLANK(.33)
/PRINT=INITIAL EXTRACTION ROTATION CORRELATION KMO
/PLOT=EIGEN
/EXTRACTION=PC
/ROTATION=VARIMAX.

(Note: The method of factor extraction used is **Principal Components analysis.**)

12.4.4   SPSS Output

TABLE 12.1

Factor Analysis Output

Correlation Matrix

		PROVO	PROTECT	MENTAL	CAUSED	SAVE	INSANE	PASSION	DEFEND	STABLE
Correlation	PROVO	1.000	.183	.144	.277	.133	.214	.288	.086	.099
	PROTECT	.183	1.000	.066	.356	.662	-.177	.023	.554	-.014
	MENTAL	.144	.066	1.000	-.107	-.031	.474	.035	-.011	.446
	CAUSED	.277	.356	-.107	1.000	.446	-.044	.095	.232	-.173
	SAVE	.133	.662	-.031	.446	1.000	-.211	.119	.655	-.119
	INSANE	.214	-.177	.474	-.044	-.211	1.000	.095	-.111	.407
	PASSION	.288	.023	.035	.095	.119	.095	1.000	.134	.060
	DEFEND	.086	.554	-.011	.232	.655	-.111	.134	1.000	-.067
	STABLE	.099	-.014	.446	-.173	-.119	.407	.060	-.067	1.000

KMO and Bartlett's Test

Kaiser–Meyer–Olkin Measure of Sampling Adequacy.		.690
Bartlett's Test of Sphericity	Approx. chi-square	930.250
	df	36
	Sig.	.000

Communalities

	Initial	Extraction
PROVO	1.000	.651
PROTECT	1.000	.744
MENTAL	1.000	.695
CAUSED	1.000	.491
SAVE	1.000	.808
INSANE	1.000	.635
PASSION	1.000	.509
DEFEND	1.000	.661
STABLE	1.000	.622

Extraction method: principal component analysis.

Total Variance Explained

Component	Initial Eigenvalues			Extraction Sums of Squared Loadings			Rotation Sums of Squared Loadings		
	Total	% of Variance	Cumulative %	Total	% of Variance	Cumulative %	Total	% of Variance	Cumulative %
1	2.663	29.591	29.591	2.663	29.591	29.591	2.452	27.244	27.244
2	1.977	21.963	51.554	1.977	21.963	51.554	1.933	21.482	48.726
3	1.176	13.069	64.623	1.176	13.069	64.623	1.431	15.897	64.623
4	.874	9.708	74.330						
5	.616	6.839	81.169						
6	.569	6.322	87.491						
7	.509	5.651	93.142						
8	.345	3.835	96.977						
9	.272	3.023	100.000						

Extraction method: principal component analysis.

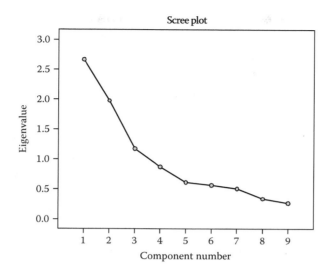

FIGURE 12.1

Component Matrix[a]

	Component		
	1	**2**	**3**
SAVE	.878		
PROTECT	.795		
DEFEND	.758		
CAUSED	.613		.331
MENTAL		.748	
INSANE	−.333	.723	
STABLE		.680	
PASSION			.606
PROVO		.513	.568

Extraction method: principal component analysis.

[a]Three components extracted.

Rotated Component Matrix[a]

	Component		
	1	2	3
SAVE	.883		
PROTECT	.861		
DEFEND	.813		
MENTAL		.830	
STABLE		.787	
INSANE		.726	
PROVO			.779
PASSION			.713
CAUSED	.455		.483

Extraction method: principal component analysis.

Rotation method: Varimax with Kaiser Normalization.

[a] Rotation converged in four iterations.

Component Transformation Matrix

Component	1	2	3
1	.919	−.288	.268
2	.153	.890	.429
3	−.362	−.354	.862

Extraction method: principal component analysis.

Rotation method: Varimax with Kaiser Normalization.

12.4.5   Results and Interpretation

12.4.5.1   Correlation Matrix

Examination of the **Correlation Matrix** (see Table 12.1) reveals fairly high correlations between the nine variables written to measure specific defense strategies. For example, the intercorrelations between the variables of PROTECT, SAVE, and DEFEND (self-defense defense strategy) are greater than 0.33. Similarly, the intercorrelations between MENTAL, INSANE, and STABLE (insanity defense strategy) are also greater than 0.33. Given the number of high intercorrelations between the defense-specific variables, the hypothesized factor model appears to be appropriate.

The **Bartlett's test of sphericity** can be used to test for the adequacy of the correlation matrix, i.e., the correlation matrix has significant correlations among at least some of the variables. If the variables are independent, the observed correlation matrix is expected to have small off-diagonal coefficients. Bartlett's test of sphericity tests the hypothesis that the correlation matrix is an identity matrix, that is, all the diagonal terms are 1 and all off-diagonal terms are 0. If the test value is large and the significance level is small (< 0.05), the hypothesis that the variables are independent can be rejected. In the present analysis, the Bartlett's test of sphericity yielded a value of 930.25 and an associated level of significance smaller than 0.001. Thus, the hypothesis that the correlation matrix is an identity matrix is rejected.

12.4.5.2 Factor Analysis Output

The **Communalities** section presents the communality of each variable (i.e., the proportion of variance in each variable accounted for by the common factors). In using the principal components method of factor extraction, it is possible to compute as many factors as there are variables. When all factors are included in the solution, all of the variance of each variable is accounted for by the common factors. Thus, the proportion of variance accounted for by the common factors, or the **communality** of a variable is 1 for all the variables.

The **Total Variance Explained** section presents the number of common factors computed, the eigenvalues associated with these factors, the percentage of total variance accounted for by each factor, and the cumulative percentage of total variance accounted for by the factors. Although nine factors have been computed, it is obvious that not all nine factors will be useful in representing the list of nine variables. In deciding how many factors to extract to represent the data, it is helpful to examine the eigenvalues associated with the factors. Using the criterion of retaining only factors with eigenvalues of 1 or greater, the first three factors will be retained for rotation. These three factors account for 29.59%, 21.96%, and 13.07% of the total variance, respectively. That is, almost 65% of the total variance is attributable to these three factors. The remaining six factors together account for only approximately 35% of the variance. Thus, a model with three factors may be adequate to represent the data. From the Scree plot, it again appears that a three-factor model should be sufficient to represent the data set.

The **Component Matrix** represents the unrotated component analysis factor matrix, and presents the correlations that relate the variables to the three extracted factors. These coefficients, called *factor loadings*, indicate how closely the variables are related to each factor. However, as the factors are unrotated (the factors were extracted on the basis of the proportion of total variance explained), significant cross-loadings have occurred. For example, the variable CAUSED has loaded highly on Factor 1 and Factor 3; the variable INSANE has loaded highly on Factor 1 and Factor 2; the variable PROVO has loaded highly on Factor 2 and Factor 3. These high cross-loadings make interpretation of the factors difficult and theoretically less meaningful.

The **Rotated Component Matrix** presents the three factors after **Varimax** (orthogonal) rotation. To subject the three factors to **Oblique** (nonorthogonal) rotation, (1) check the **Direct Oblimin** field in the **Factor Analysis: Rotation** window, or (2) substitute the word **VARIMAX** with **OBLIMIN** in the ROTATION subcommand in the SPSS Syntax Method in Subsection 12.4.3. The OBLIMIN rotation output is presented in Table 12.2.

TABLE 12.2

Oblique (OBLIMIN) Rotation Output

Pattern Matrix[a]

	Component		
	1	2	3
SAVE	.878		
PROTECT	.876		
DEFEND	.825		
MENTAL		.844	
STABLE		.794	
INSANE		.704	
PROVO			.776
PASSION			.724
CAUSED	.394		.459

Extraction method: principal component analysis.

Rotation method: Oblimin with Kaiser Normalization.

[a.] Rotation converged in five iterations.

Component Correlation Matrix

Component	1	2	3
1	1.000	−140	.177
2	−.140	1.000	6.465E-02
3	.177	6.465E-02	1.000

Extraction method: principal component analysis.

Rotation method: Oblimin with Kaiser Normalization.

As there is no overwhelming theoretical reason to employ one rotation method over another, the decision to interpret either the **varimax** rotated matrix or the **oblimin** matrix depends on the magnitude of the factor correlations presented in the **Component Correlation Matrix** (in Table 12.2). Examination of the factor correlations indicates that the three factors are not strongly correlated (all coefficients are less than 0.20), which suggests that the varimax (orthogonal) matrix should be interpreted. However, should the decision be made to interpret the oblimin rotated matrix, then a further decision must be made to interpret either the **Pattern Matrix** or the **Structure Matrix**. The structure matrix presents the correlations between variables and factors, but these may be confounded by correlations between the factors. The pattern matrix shows the uncontaminated correlations between variables and factors and is generally used for interpreting factors.

Examination of the factor loadings presented in the **Varimax Rotated Component Matrix** (Table 12.1) shows that eight of the nine variables loaded highly on the three factors representing the three defense strategies of self-defense, insanity, and provocation. One variable, CAUSED, cross-loaded significantly across Factor 1 and Factor 3. Convention suggests three possible ways of handling significant cross-loadings.

1. If the matrix indicates many significant cross-loadings, this may suggest further commonality between the cross-loaded variables and the factors. The researcher may decide to rerun factor analysis, stipulating a smaller number of factors to be extracted.
2. Examine the wording of the cross-loaded variables, and based on their face-validity, assign them to the factors that they are most conceptually/logically representative of.
3. Delete all cross-loaded variables. This will result in "clean" factors and will make interpretation of the factors that much easier. This method works best when there are only few significant cross-loadings.

In the present example, the cross-loaded variable of CAUSED appears to be more conceptually relevant to Factor 3 (provocation defense) than to Factor 1 (self-defense defense). Thus, the decision may be made to retain this variable to represent Factor 3. Alternatively, the researcher may decide to delete this variable. In any case, no further analysis (rotation) is required as the factor structure is clearly consistent with the hypothesized three-factor model.

In summary, it can be concluded that factor analysis has identified three factors from the list of nine variables. In the main, these factors are represented by the specific statements written to reflect the three defense strategies of self-defense, insanity, and provocation.

12.5 Factor Analysis: Example 2

This example demonstrates the use of factor analysis in deriving meaningful factors using multiple runs. A study was designed to identify the motives for the maintenance of smoking behavior and its possible cessation (Ho, 1989). Twenty five statements were written to represent these motives. Each statement was rated on a five-point scale with high scores indicating strong agreement with that motive as a reason for smoking. A total of 91 smokers provided responses to these 25 statements. Factor analysis (with principal components extraction, followed by varimax rotation) was employed to investigate the factor structure of this 25-item smoking inventory. The 25 statements written to reflect smoking motives are listed in Table 12.3.

TABLE 12.3

Reasons for Smoking

1	I find smoking enjoyable.
2	When I feel stressed, tense, or nervous, I light up a cigarette.
3	Smoking lowers my appetite and therefore keeps my weight down.
4	Lighting up a cigarette is a habit to me.
5	I smoke cigarettes to relieve boredom.
6	Smoking gives me something to do with my hands.
7	I feel secure when I am smoking.
8	I enjoy lighting up after pleasurable experiences, e.g., after a good meal.
9	Smoking relaxes me.
10	Smoking gives me a lift.
11	Smoking allows me to be "part of a crowd."
12	I smoke because I am addicted to cigarettes.
13	I smoke because members of my family smoke.
14	Smoking is a means of socializing.
15	Smoking helps me to concentrate when I am working.
16	I smoke because most of my friends smoke.
17	I smoke because it makes me feel confident.
18	Smoking makes me feel sophisticated and glamorous.
19	I smoke as an act of defiance.
20	I smoke because I find it difficult to quit.
21	I enjoy the taste of cigarettes.
22	I find smoking pleasurable.
23	I smoke to annoy nonsmokers.
24	The health statistics regarding smoking cigarettes and health problems don't bother me, as they are highly exaggerated anyway.
25	I am willing to live with my health problems that my smoking may cause me.

12.5.1   Data Entry Format

The data set has been saved under the name: **SMOKE.SAV**.

Variables	Columns	Code
s1 to s25	1–25	1 = strong agree, 5 = strongly disagree

12.5.2   Windows Method (First Run)

1. From the menu bar, click **Analyze**, then **Data Reduction**, and then **Factor**. The following **Factor Analysis** window will open.

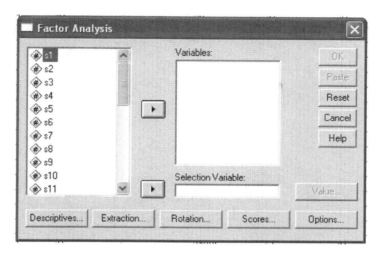

2. Transfer the 25 variables of **S1 to S25** to the **Variables** field by clicking these variables (highlighting) and then clicking 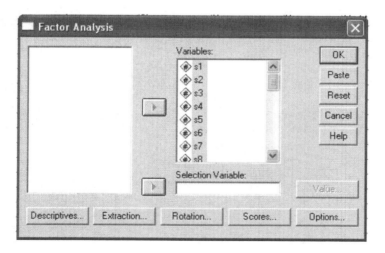 .

3. To test that the correlation matrix, generated from the 25 variables, has sufficient correlations to justify the application of factor analysis, click **Descriptives...** . The following **Factor Analysis: Descriptives** window will open. Check the and **KMO and Bartlett's test of sphericity** field, and then click **Continue** .

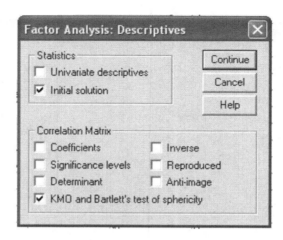

4. When the **Factor Analysis** window opens, click . This will open the **Factor Analysis: Extraction** window. In the **Method** drop-down list, choose **Principal components** as the extraction method. Ensure that the **Correlation matrix** field is checked. In the **Eigenvalues over** field, accept the default value of **1**. Leave the **Number of factors** field blank (i.e., allow principal components analysis to extract as many factors as there are with eigenvalues greater than 1). To obtain a **Scree plot** of the number of factors extracted, check the **Scree plot** field. Click **Continue**.

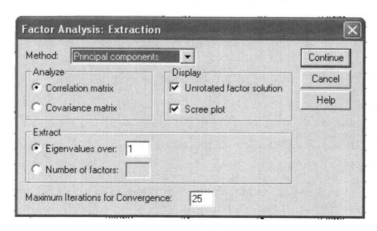

5. When the **Factor Analysis** window opens, click **Rotation...**. This will open the **Factor Analysis: Rotation** window. To subject the

extracted factors to **Varimax** rotation, check the **Varimax** field. Click **Continue** .

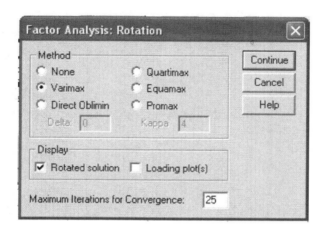

6. When the **Factor Analysis** window opens, click **Options...** . This will open the **Factor Analysis: Options** window. If the data set has **missing values**, the researcher can choose one of the three methods offered to deal with the missing values: (1) **Exclude cases listwise**, (2) **Exclude cases pairwise**, and (3) **Replace with mean**. Under **Coefficient Display Format**, check the **Sorted by size** cell. This procedure will present the factor loadings (correlation coefficients) in a descending order of magnitude format in the output. Check the **Suppress absolute values less than** field, and then type the coefficient of **0.33** in the field next to it. This procedure will suppress the presentation of any factor loadings with values less than 0.33 in the output. Click **Continue** .

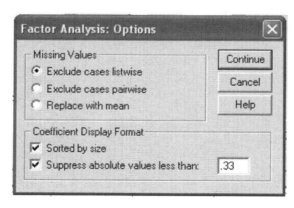

7. When the **Factor Analysis** window opens, click 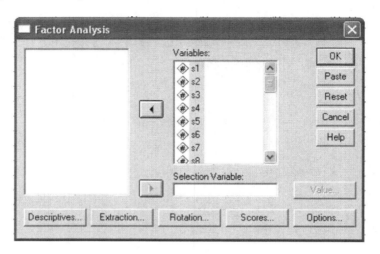 OK to complete the analysis. See Table 12.4 for the results.

12.5.3   SPSS Syntax Method (First Run)

FACTOR VARIABLES=S1 TO S25
/FORMAT=SORT BLANK(.33)
/PRINT=INITIAL EXTRACTION ROTATION KMO
/PLOT=EIGEN
/EXTRACTION=PC
/ROTATION=VARIMAX.

12.5.4   SPSS Output

TABLE 12.4

Factor Analysis Output

KMO and Bartlett's Test		
Kaiser–Meyer–Olkin Measure of Sampling Adequacy		.687
Bartlett's Test of Sphericity	Approx. Chi-Square	876.290
	df	300
	Sig.	.000

	Communalities	
	Initial	Extraction
S1	1.000	.747
S2	1.000	.636
S3	1.000	.399
S4	1.000	.535
S5	1.000	.751
S6	1.000	.794
S7	1.000	.698
S8	1.000	.687
S9	1.000	.658
S10	1.000	.723
S 11	1.000	.722
S12	1.000	.727
S13	1.000	.610
S14	1.000	.703
S15	1.000	.593
S16	1.000	.795
S17	1.000	.698
S18	1.000	.740
S19	1.000	.648
S20	1.000	.695
S21	1.000	.578
S22	1.000	.722
S23	1.000	.558
S24	1.000	.716
S25	1.000	.403

Extraction method: principal component analysis.

Total Variance Explained

Component	Initial Eigenvalues			Extraction Sums of Squared Loadings			Rotation Sums of Squared Loadings		
	Total	% of Variance	Cumulative %	Total	% of Variance	Cumulative %	Total	% of Variance	Cumulative %
1	5.315	21.258	21.258	5.315	21.258	21.258	4.199	16.797	16.797
2	2.891	11.565	32.823	2.891	11.565	32.823	2.550	10.201	26.998
3	2.636	10.544	43.367	2.636	10.544	43.367	2.169	8.675	35.672
4	1.666	6.662	50.029	1.666	6.662	50.029	2.165	8.660	44.332
5	1.512	6.049	56.078	1.512	6.049	56.078	2.022	8.089	52.421
6	1.324	5.296	61.375	1.324	5.296	61.375	1.979	7.916	60.338
7	1.190	4.760	66.134	1.190	4.760	66.134	1.449	5.797	66.134
8	.937	3.748	69.882						
9	.897	3.590	73.471						
10	.806	3.224	76.696						
11	.732	2.930	79.625						
12	.689	2.755	82.381						
13	.571	2.286	84.666						
14	.561	2.244	86.910						
15	.514	2.057	88.968						
16	.451	1.805	90.773						
17	.370	1.479	92.252						
18	.354	1.414	93.667						
19	.322	1.287	94.954						
20	.307	1.228	96.182						
21	.258	1.034	97.216						
22	.232	.929	98.145						
23	.180	.722	98.866						
24	.166	.666	99.532						
25	.117	.468	100.000						

Extraction method: principal component analysis.

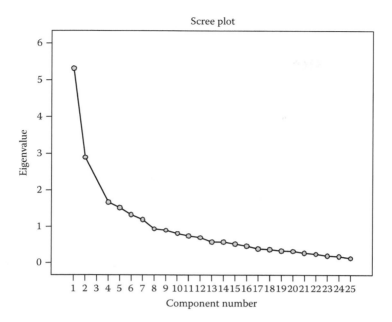

FIGURE 12.2

Component Matrix[a]

	Component						
	1	2	3	4	5	6	7
SI8	.772						
S17	.769						
S16	.764						
S11	.751						
S14	.680						
S7	.676	.442					
S10	.606					-.339	
S13	.576			.357		.347	
S5	.493	.355		.409			-.371
S3	.388		-.337				
S12		.734			.357		
S20		.607					.501
S4		.584		.391			
S2		.547			-.459		
S24	.431	-.448		.377		.357	
S1			.844				
S22			.809				
S21			.623				
S23	.404			.520			
S6	.418	.363		.507			
S19	.361				.552		
S9	.355		.336	-.360	-.440		
S8		.345	.476			.511	
S25							.395
S15					.350	.367	-.384

Extraction method: principal component analysis.

[a]Seven components extracted.

Rotated Component Matrix[a]

	Component					
	1	2	3	4	5	7
S14	.802					
S11	.780					
S18	.770					
S17	.748					
S16	.744		.401			
S10	.592	.441				
519	.485				−.477	.390
S11		.842				
S22		.830				
S21		.691				
S3		−.357				
S24			.791			
S13			.680			
S23			.643			
S20				.809		
S12				.788		
S4				.594	.347	
S2					.737	
S9					.734	
S7	.460				.502	.357
S5					.812	
S6					.808	
S15						.724
S25						−.536
S8		.383		.346		.493

Extraction method: principal component analysis.

Rotation method: Varimax with Kaiser Normalization.

[a]Rotation converged in nine iterations.

Component Transformation Matrix

Component	1		3	4	5	6	7
1	.830	.099	.379	.066	.218	.303	.116
2	−.173	−.055	−.316	.727	.427	.343	.195
3	−.144	.945	.038	−032	.255	−.139	.006
4	−.371	.112	.581	.182	−383	.552	−.167
5	.188	.199	−.142	.346	−.683	−.190	.532
6	−.250	−.206	.603	.078	.297	−.427	.506
7	.158	−.017	.185	.554	−.063	−.498	−.618

Extraction method: principal component analysis.

Rotation method: Varimax with Kaiser Normalization.

12.5.5   Results and Interpretation

12.5.5.1   Correlation Matrix

The **Bartlett's Test of Sphericity** (see Table 12.4) tests the adequacy of the correlation matrix, and yielded a value of 876.29 and an associated level of significance smaller than 0.001. Thus, the hypothesis that the correlation matrix is an identity matrix can be rejected, i.e., the correlation matrix has significant correlations among at least some of the variables.

12.5.5.2   Factor Analysis Output

The **Total Variance Explained** section presents the number of common factors extracted, the eigenvalues associated with these factors, the percentage of total variance accounted for by each factor, and the cumulative percentage of total variance accounted for by the factors. Using the criterion of retaining only factors with eigenvalues of 1 or greater, seven factors were retained for rotation. These seven factors accounted for 21.26%, 11.56%, 10.54%, 6.66%, 6.05%, 5.30%, and 4.76% of the total variance, respectively, for a total of 66.13%. The scree plot, however, suggests a four-factor solution.

The **Rotated Component Matrix** presents the seven factors after varimax rotation. To identify what these factors represent, it would be necessary to consider what items loaded on each of the seven factors. The clustering of the items in each factor and their wording offer the best clue as to the meaning of that factor. For example, eight items loaded on Factor 1. An inspection of these items (see Table 12.3) clearly shows that the majority of these items reflect a **social** motive for smoking (e.g., smoking is a means of socializing; smoking allows me to be part of a crowd; I smoke because most of my friends smoke, etc.). Factor 2 contains five items that clearly reflect the **pleasure** that a smoker gains from smoking (e.g., I find smoking enjoyable; I find smoking pleasurable; I enjoy the taste of cigarettes, etc). Factors 4, 5, and 6 contain items that appear to reflect two related motives — **addiction** and **habit** (e.g., I smoke because I find it difficult to quit; I smoke because I am addicted to cigarettes; lighting up is a habit to me; smoking gives me something to do with my hands, etc.). The two remaining factors, Factor 3 and Factor 7, contain items that do not "hang" together conceptually, and as such, are not easily interpretable. In fact, some of the items that load on these two factors appear to overlap in meaning with other factors. For example, item s13 (I smoke because members of my family smoke) in Factor 3 appears to reflect a social motive, and thus overlaps in meaning with Factor 1. Similarly, item s8 (I enjoy lighting up after pleasurable experiences) in Factor 7 appears to overlap in meaning with Factor 2 (pleasure motive). The commonality in meaning of some of these factors suggests that a number of factors can be combined. The combination of factors is purely a subjective decision, aimed at reducing the number of extracted factors to a smaller, more manageable, and ultimately more meaningful set of factors. Given that the present factor structure appears to be represented by three dimensions of smoking motives (**Social**, **Pleasure**, and **Addiction/Habit**), it was decided to rerun Factor Analysis, stipulating the extraction of only three factors.

12.5.6   Windows Method (Second Run)

1. From the menu bar, click **Analyze**, then **Data Reduction**, and then **Factor**. The following **Factor Analysis** window will open.

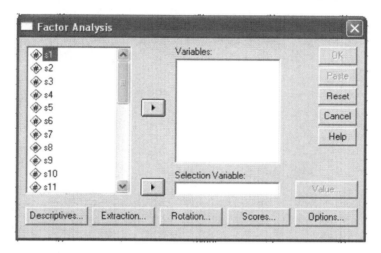

2. Transfer the 25 variables of **S1 to S25** to the **Variables** cell by clicking these variables (highlighting) and then clicking ▶ .

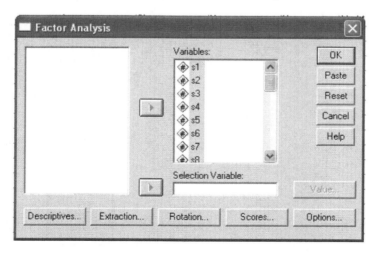

3. Click to open the **Factor Analysis: Extraction** window. In the **Method** drop-down list, choose **Principal components** as the extraction method. To extract only three factors from the correlation matrix, check the **Number of factors** field, and then type **3** in the field next to it. This procedure will override the default extraction

of all factors with eigenvalues greater than 1. Click Continue .

4. When the **Factor Analysis** window opens, click Rotation... . This will open the **Factor Analysis: Rotation** window. To subject the extracted factors to **Varimax** rotation, check the **Varimax** cell. Click Continue .

5. When the **Factor Analysis** window opens, click **Options...** . This will open the **Factor Analysis: Options** window. Under **Coefficient Display Format**, check the **Sorted by size** field. Check the **Suppress absolute values less than** field, and then type the coefficient of **0.33** in the field next to it. Click **Continue** .

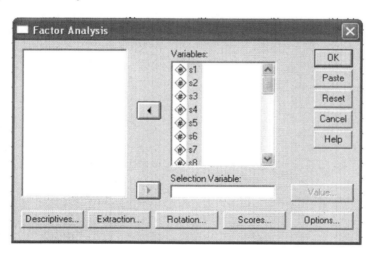

6. When the **Factor Analysis** window opens, click **OK** to complete the analysis. See Table 12.5 for the results.

12.5.7   SPSS Syntax Method (Second Run)

FACTOR VARIABLES=S1 TO S25
/FORMAT=SORT BLANK(.33)
/CRITERIA=FACTOR(3)
/EXTRACTION=PC
/ROTATION=VARIMAX.

12.5.8   SPSS Output

TABLE 12.5

Three-Factor Structure Output

Total Variance Explained

Component	Initial Eigenvalues			Extraction Sums of Squared Loadings			Rotation Sums of Squared Loadings		
	Total	% of Variance	Cumulative %	Total	% of Variance	Cumulative %	Total	% of Variance	Cumulative %
1	5.315	21.258	21.258	5.315	21.258	21.258	4.953	19.814	19.814
2	2.891	11.565	32.823	2.891	11.565	32.823	3.163	12.651	32.465
3	2.636	10.544	43.367	2.636	10.544	43.367	2.725	10.902	43.367
4	1.666	6.662	50.029						
5	1.512	6.049	56.078						
6	1.324	5.296	61.375						
7	1.190	4.760	66.134						
8	.937	3.748	69.882						
9	.897	3.590	73.471						
10	.806	3.224	76.696						
11	.732	2.930	79.625						
12	.689	2.755	82.381						
13	.571	2.286	84.666						
14	.561	2.244	86.910						
15	.514	2.057	88.968						
16	.451	1.805	90.773						
17	.370	1.479	92.252						
18	.354	1.414	93.667						
19	.322	1.287	94.954						
20	.307	1.228	96.182						
21	.258	1.034	97.216						
22	.232	.929	98.145						
23	.180	.722	98.866						
24	.166	.666	99.532						
25	.117	.468	100.000						

Extraction method: principal component analysis.

Component Matrix[a]

	Component		
	1	2	3
S18	.772		
S17	.769		
S16	.764		
S11	.751		
S14	.680		
S7	.676	.442	
S10	.606		
S13	.576		
S5	.493	.355	
S6	.418	.363	
S23	.404		
S3	.388		−.337
S19	.361		
S9	.355		.336
S15			
S12		.734	
S20		.607	
S4		.584	
S2		.547	
S24	.431	−.448	
S1			.844
S22			.809
S21			.623
S8		.345	.476
S25			

Extraction method: principal component analysis.

[a]Three components extracted.

Rotated Component Matrix[a]
Component

	1	2	3
S11	.796		
S16	.790		
S18	.783		
S17	.750		
S14	.721		
S13	.567		
S24	.525		
S23	.486		
S19	.463		
S10	.439	.370	.421
S12		.685	
S7	.478	.643	
S20		.607	
S4		.561	
S2		.555	
S5	.350	.502	
S6		.491	
S3		.421	
S15			
S 1			.850
S22			.827
S21			.629
S8		.356	.487
S9			.392
S25			

Extraction method: principal component analysis.

Rotation method: Varimax with Kaiser Normalization.

[a]Rotation converged in four iterations.

12.5.9   Results and Interpretation

The results presented in the **Total Variance Explained** section (see Table 12.5) are identical to those obtained in the first run (Table 12.4). This is not surprising as the same extraction method (principal components analysis) was applied to the same 25 items. Thus, the same seven factors were extracted, accounting for a combined 66.13% of the total variance.

The **Rotated Component Matrix** presents only three rotated factors as stipulated in both the SPSS windows and syntax file methods. The rotated factor structure shows a number of cross-loaded items (s10, s7, s5, and s8) that were deleted prior to interpretation. Deletion of cross-loaded items serves to clarify the factors and makes their interpretation easier. Factor 1 contains nine items that clearly reflect the social motive for smoking, and was thus labeled **SOCIAL**. Factor 2 contains six items that reflect addiction and habit as motives for smoking, and was labeled **ADDICTION/HABIT**. Factor 3 contains four items that reflect the pleasure gained from smoking, and was labeled **PLEASURE**. This three-factor model represents the combination of the seven original factors, and appears to reflect adequately the underlying factor structure of the 25-item smoking inventory.

13

Reliability

13.1 Aim

The reliability of a measuring instrument is defined as its ability to consistently measure the phenomenon it is designed to measure. Reliability, therefore, refers to **test consistency**. The importance of reliability lies in the fact that it is a prerequisite for the validity of a test. Simply put, for the validity of a measuring instrument to be supported, it must be demonstrably reliable. Any measuring instrument that does not reflect some attribute consistently has little chance of being considered a valid measure of that attribute.

Several methods exist for determining the reliability of a measuring instrument. These methods may be divided into two broad categories: **external consistency** procedures, and **internal consistency** procedures.

13.1.1 External Consistency Procedures

External consistency procedures utilize cumulative test results against themselves as a means of verifying the reliability of the measure. Two major methods of determining the reliability of a test by external consistency are:

1. **Test-retest:** Test results for a group of people are compared in two different time periods.
2. **Parallel forms of the same test:** Two sets of results from equivalent but different tests are compared.

13.1.2 Internal Consistency Procedures

Internal consistency refers to the extent to which the items in a test measure the same construct. Items that measure the same phenomenon should logically cling/hang together in some consistent manner. Examining the internal consistency of the test enables the researcher to determine which items are not consistent with the test in measuring the phenomenon under investigation. The object is to remove the inconsistent items and improve the internal consistency of the test. An internally consistent test increases the chances of the test being reliable. Three major methods of determining the reliability of a test by internal consistency are:

1. **Split-half technique:** This method correlates one half of the test items with the other half. The higher the correlation, the more internally consistent the measure. The **Spearman-Brown** formula is widely used in determining reliability by the split-half method.

2. **Cronbach's alpha:** This is a single correlation coefficient that is an estimate of the average of all the correlation coefficients of the items within a test. If alpha is high (0.80 or higher), then this suggests that all of the items are reliable and the entire test is internally consistent. If alpha is low, then at least one of the items is unreliable, and must be identified via item analysis procedure.

3. **Item analysis:** This procedure represents a refinement of test reliability by identifying "problem" items in the test, i.e., those items that yield low correlations with the sum of the scores on the remaining items. Rejecting those items that are inconsistent with the rest (and retaining those items with the highest average intercorrelations) will increase the internal consistency of the measuring instrument. Item analysis is achieved through **Item-Total correlation** procedure.

13.2   Example — Reliability

In Example 1 in Chapter 12 on factor analysis, three factors were identified to reflect various defense strategies for use in trials of battered women who killed their abusive spouses. These three factors (extracted from nine variables) represent the defense of **self-defense**, **insanity**, and **provocation**. Before these factors can be computed and used in subsequent analyses, the internal consistency of each factor should be tested to ensure the reliability of the factors. For this example, only the reliability of the **self-defense** factor will be tested. The data set to be analyzed is the same as the one used in Example 1 on factor analysis (**DOMES.SAV**).

13.2.1   Windows Method

1. From the menu bar, click **Analyze**, then **Scale**, and then **Reliability Analysis**. The following **Reliability Analysis** window will open.

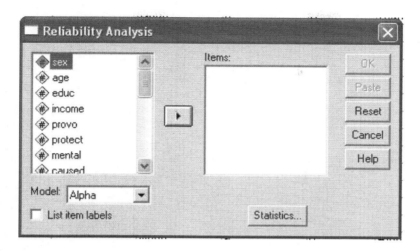

2. Because the **self-defense** factor incorporates the measurement items of **PROTECT, SAVE,** and **DEFEND,** transfer these three items to the **Items** field by clicking these items (highlighting) and then clicking

   . In the **Model** field select **Alpha** from the drop-down list. **Alpha (Cronbach's Alpha)** is the default for testing the overall internal consistency of a factor.

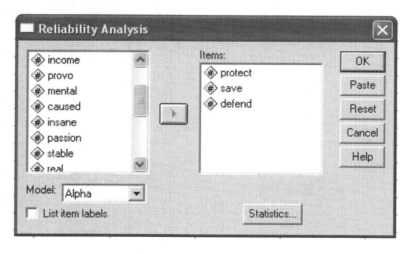

3. Click Statistics... to open the **Reliability Analysis: Statistics** window. In the **Descriptives for** section, check the **Scale** and the **Scale if item deleted** fields. Click Continue.

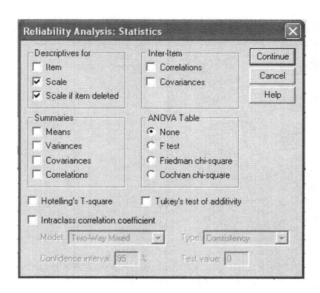

4. When the **Reliability Analysis** window opens, click [OK] to complete the analysis. See Table 13.1 for the results.

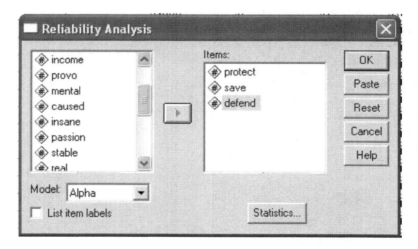

13.2.2   SPSS Syntax Method

RELIABILITY VARIABLES=PROVO TO STABLE
/SCALE(SELF)=PROTECT SAVE DEFEND
/SUMMARY=TOTAL.

13.2.3 SPSS Output

TABLE 13.1

Reliability Output

Case-Processing Summary

		N	%
Cases	Valid	394	99.2
	Excluded[a]	3	.8
	Total	397	100.0

[a]Listwise deletion based on all variables in the procedure.

Reliability Statistics

Cronbach's Alpha	N of Items
.831	3

Item-Total Statistics

	Scale Mean if Item Deleted	Scale Variance if Item Deleted	Corrected Item-Total Correlation	Cronbach's Alpha if Item Deleted
protect	10.29	13.356	.666	.791
save	10.00	12.995	.747	.712
defend	10.81	13.113	.662	.796

Scale Statistics

Mean	Variance	Std. Deviation	N of Items
15.55	27.296	5.225	3

13.2.4 Results and Interpretation

Of the total sample of 397 cases, 394 were processed in this analysis. Three cases were excluded from the analysis due to missing values (listwise exclusion of cases).

Cronbach's Alpha is 0.83, which indicates high overall internal consistency among the three items representing the self-defense factor. The **Corrected Item-Total Correlation** shows the correlation (consistency) between each item and the sum of the remaining items. In deciding which item to retain or delete, the 0.33 criterion can be used (an item-total correlation of 0.33 indicates that approximately 10% of the variance in the scale is accounted for by that item). Based on this criterion, all three items will be retained. Indeed, deleting any of the three items will reduce the overall reliability of the scale, as indicated by the column **Cronbach's Alpha if Item Deleted**.

14

Multiple Regression

14.1 Aim

Multiple regression is a statistical technique through which one can analyze the relationship between a dependent or criterion variable and a set of independent or predictor variables. As a statistical tool, multiple regression is frequently used to achieve three objectives.

1. To find the best prediction equation for a set of variables; i.e., given X and Y (the predictors), what is Z (the criterion variable)?
2. To control for confounding factors to evaluate the contribution of a specific variable or set of variables, i.e., identifying independent relationships.
3. To find structural relationships and provide explanations for seemingly complex multivariate relationships, such as is done in path analysis.

14.2 Types of Multiple Regression Method

There are three major types of multiple regression technique: **standard multiple regression, hierarchical regression**, and **statistical (stepwise) regression**. They differ in terms of how the overlapping variability due to correlated independent variables is handled, and who determines the order of entry of independent variables into the equation (Tabachnick & Fidell, 1989).

14.2.1 Standard Multiple Regression

For this regression model, all the study's independent variables are entered into the regression equation at once. Each independent variable is then assessed in terms of the unique amount of variance it accounts for. The disadvantage of the standard regression model is that it is possible for an independent variable to be strongly related to the dependent variable, and yet be considered an unimportant predictor, if its unique contribution in explaining the dependent variable is small.

14.2.2   Hierarchical Multiple Regression

This regression model is most flexible as it allows the researcher to determine the order of entry of the independent variables into the regression equation. Each independent variable is assessed at its own point of entry in terms of the additional explanatory power it contributes to the equation. The order of entry is normally dictated by logical or theoretical considerations. For example, based on theoretical reasons, a researcher may decide that two specific independent variables (from a set of independent variables) will be the strongest predictors of the dependent variable. Thus, these two independent variables will be accorded priority of entry, and their total explanatory power (in terms of the total amount of variance explained) evaluated. Then the less important independent variables are entered and evaluated in terms of what they add to the explanation above and beyond that afforded by the first two independent variables. It is also possible to take the opposite tack in which less important independent variables are entered into the equation first, to cull away "nuisance" variance. Then the important set of independent variables is entered and evaluated in terms of what it adds to the explanation of the dependent variable.

14.2.3   Statistical (Stepwise) Regression

For this statistical regression model, the order of entry of predictor variables is based solely on statistical criteria. Variables that correlate most strongly with the dependent variable will be afforded priority of entry, with no reference to theoretical considerations. The disadvantage of this type of regression is that the statistical criteria used for determining priority of entry may be specific to the sample at hand. For another sample, the computed statistical criteria may be different, resulting in a different order of entry for the same variables. The statistical regression model is used primarily in exploratory work, in which the researcher is unsure about the relative predictive power of the study's independent variables.

Statistical regression can be accomplished through one of three methods: **forward selection**, **backward deletion**, and **stepwise regression**.

In **forward selection**, the variables are evaluated against a set of statistical criteria, and if they meet these criteria, are afforded priority of entry based on their relative correlations with the dependent variable. The variable that correlates most strongly with the dependent variable gets entered into the equation first, and once in the equation, it remains in the equation.

In **backward deletion**, the equation starts out with all the independent variables entered. Each variable is then evaluated one at a time, in terms of its contribution to the regression equation. Those variables that do not contribute significantly are deleted.

In **stepwise regression**, variables are evaluated for entry into the equation under both forward selection and backward deletion criteria. That is, variables are entered one at a time if they meet the statistical criteria, but they may also be deleted at any step where they no longer contribute significantly to the regression model.

14.3 Checklist of Requirements

* The size of the sample has a direct impact on the statistical power of the significance testing in multiple regression. **Power** in multiple regression refers to the probability of detecting as statistically significant a specific level of R-square, or a regression coefficient at a specified significance level and a specific sample size (Hair et al., 1995). Thus, for a desired level of power and with a specified number of independent variables, a certain sample size will be required to detect a significant R-square at a specified significance level (see Cohen and Cohen, 1983, for sample size calculations). As a rule of thumb, there should be at least 20 times more cases than independent variables. That is, if a study incorporates five independent variables, there should be at least 100 cases.

* The measurement of the variables can be either continuous (metric) or dichotomous (nonmetric). When the dependent variable is dichotomous (coded 0-1), discriminant analysis is appropriate. When the independent variables are discrete, with more than two categories, they must be converted into a set of dichotomous variables by **dummy variable coding**. A dummy variable is a dichotomous variable that represents one category of a nonmetric independent variable. Any nonmetric variable with K categories can be represented as K-1 dummy variables, i.e., one dummy variable for each degree of freedom. For example, an ethnicity variable may originally be coded as a discrete variable with 1 for Australian, 2 for Chinese, 3 for European, and 4 for others. The variable can be converted into a set of three dummy variables (Australian vs. non-Australian, Chinese vs. non-Chinese, and European vs. non-European), one for each degree of freedom. This new set of dummy variables can be entered into the regression equation.

14.4 Assumptions

- **Linearity:** As regression analysis is based on the concept of correlation, the linearity of the relationship between dependent and independent variables is crucial. Linearity can easily be examined by residual plots. For nonlinear relationships, corrective action to accommodate the curvilinear effects of one or more independent variables can be taken to increase both the predictive accuracy of the model and the validity of the estimated coefficients.

- **Homoscedasticity:** The assumption of equal variances between pairs of variables. Violation of this assumption can be detected by either residual plots or simple statistical tests. SPSS provides the **Levene Test for Homogeneity of Variance**, which measures the equality of variances for a single pair of variables.

- **Independence of error terms:** In regression, it is assumed that the predicted value is not related to any other prediction; i.e., each predicted value is independent. Violation of this assumption can be detected by plotting the residuals against sequence of cases. If the residuals are independent, the pattern should appear random. Violations will be indicated by a consistent pattern in the residuals. SPSS provides the **Durbin-Watson** statistic as a test for serial correlation of adjacent error terms, and, if significant, indicates nonindependence of errors.

- **Normality of the error distribution:** It is assumed that errors of prediction (differences between the obtained and predicted dependent variable scores) are normally distributed. Violation of this assumption can be detected by constructing a histogram of residuals, with a visual check to see whether the distribution approximates the normal distribution.

14.5 Multicollinearity

Multicollinearity refers to the situation in which the independent/predictor variables are highly correlated. When independent variables are multicollinear, there is "overlap" or sharing of predictive power. This may lead to the paradoxical effect, whereby the regression model fits the data well, but none of the predictor variables has a significant impact in predicting the dependent variable. This is because when the predictor variables are highly correlated, they share essentially the same information. Thus, together, they may explain a great deal of the dependent variable, but may not *individually* contribute significantly to the model. Thus, the impact of multicollinearity is to reduce any individual independent variable's predictive power by the extent to which it is associated with the other independent variables. That is, none of the predictor variables may contribute uniquely and significantly to the prediction model after the others are included.

Checking for multicollinearity: In SPSS, it is possible to request the display of "Tolerance" and "VIF" values for each predictor as a check for multicollinearity. A tolerance value can be described in this way. From a set of three predictor variables, use X_1 as a dependent variable, and X_2 and X_3 as predictors. Compute the R^2 (the proportion of variance that X_2 and X_3 explain in X_1), and then take $1 - R^2$. Thus, the tolerance value is an indication of the percentage of variance in the predictor that cannot be accounted for by the other predictors. Hence, very small values indicate "overlap" or sharing of predictive power (i.e., the predictor is redundant). Values that are less than 0.10 may merit further investigation. The VIF, which stands for *variance inflation factor*, is computed as "1/tolerance," and it is suggested that predictor variables whose VIF values are greater than 10 may merit further investigation.

Most multiple regression programs have default values for tolerance that will not admit multicollinear variables. Another way to handle the problem of multicollinearity is to either retain only one variable to represent the multicollinear variables, or combine the highly correlated variables into a single composite variable.

14.6 Example 1 — Prediction Equation and Identification of Independent Relationships (Forward Entry of Predictor Variables)

Using the data set employed in Example 1 in Chapter 12 (**DOMES.SAV**), assume that the researcher is interested in predicting the level of responsibility attributed to the battered woman for her fatal action, from the scores obtained from the three defense strategies of self-defense, provocation, and insanity. Specifically, the researcher is interested in predicting the level of responsibility attributed by a subject who strongly believes that the battered woman's action was motivated by self-defense and provocation (a score of 8 on both scales) and not by an impaired mental state (a score of 1 on the insanity scale). Attribution of responsibility is measured on an 8-point scale, with 1 = not at all responsible to 8 = entirely responsible. In addition to predicting the level of responsibility attributed to the battered woman, the researcher also wanted to identify the independent relationships between the three defense strategies and the dependent variable of responsibility attribution (coded **RESPON**; see Subsection 12.4.1). Multiple regression will be used to achieve these two objectives. Prior to multiple regression, the three defense strategies will be computed from the variables identified through factor analysis (Chapter 12). These three defense strategies are coded:

PROVOKE: provocation defense

SELFDEF: self-defense defense

INSANITY: insanity defense

14.6.1   Windows Method

Step 1. Compute the three variables of **PROVOKE, SELFDEF**, and **INSANITY**.

1. To compute the three variables of **PROVOKE, SELFDEF**, and **INSANITY**, click **Transform** and then **Compute** from the menu bar. The following **Compute Variable** window will open.

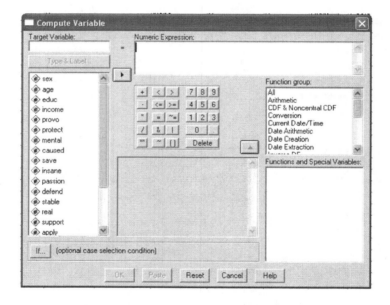

2. To compute the first variable of **PROVOKE**, type the name **PRO-VOKE** in the **Target Variable** field. As this variable is computed as the mean of the three variables of **PROVO, PASSION**, and **CAUSED** (identified through factor analysis in Chapter 12), type **MEAN** in the **Numeric Expression** field. Next, in the **Type & Label...** field, click (highlight) the three variables of **PROVO, PASSION**, and

    CAUSED, and then click to transfer these three variables to the **Numeric Expression** field as shown in the **Compute Variable** window. Click [OK] to compute the variable of **PROVOKE**.

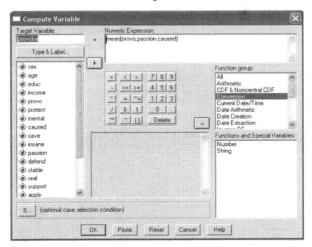

3. Repeat step 2 to compute the variables of **SELFDEF** and **INSANITY**.

Step 2. Predicting the level of responsibility from the three defense strategies of **PROVOKE**, **SELFDEF**, and **INSANITY**.

1. To predict the level of responsibility attributed to the battered woman's action from the three defense strategies of **PROVOKE**, **SELFDEF**, and **INSANITY**, click **Analyze**, then **Regression**, and then **Linear...** from the menu bar. The following **Linear Regression** window will open. Notice that the list of variables now contains the newly created variables of **PROVOKE**, **SELFDEF**, and **INSANITY**.

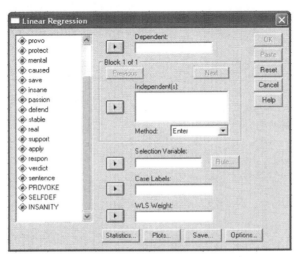

2. Click (highlight) the variable of **RESPON** and then click 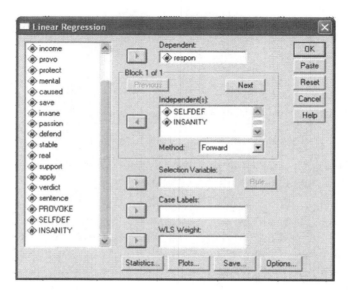 to transfer this variable to the **Dependent** field. Next, click (highlight) the three newly created variables of **PROVOKE**, **SELFDEF**, and

   INSANITY, and then click ▶ to transfer these variables to the **Independent(s)** field. In the **Method** field, select **FORWARD** from the drop-down list as the method of entry for the three independent (predictor) variables into the prediction equation.

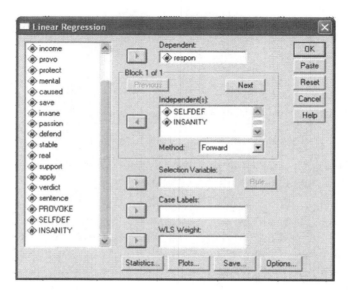

3. Click **Statistics...** to open the **Linear Regression: Statistics** window. Check the fields to obtain the statistics required. For this example, check the fields for **Estimates, Confidence intervals, Model fit, R squared change**, and **Collinearity diagnostics**. Click **Continue** when finished.

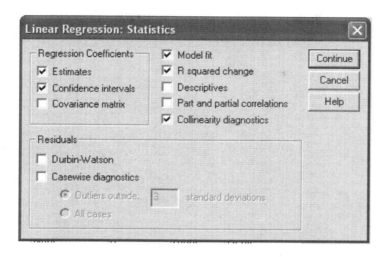

4. When the **Linear Regression** window opens, click to open the **Linear Regression: Options** window. Ensure that both the **Use probability of F** and the **Include constant in equation** fields are checked. Under **Missing Values**, check the **Exclude cases listwise** field.

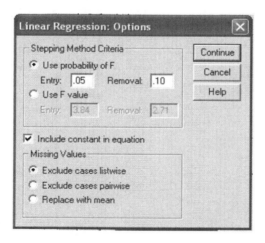

Click **Continue** to return to the **Linear Regression** window.

5. When the **Linear Regression** window opens, click **OK** to complete the analysis. See Table 14.1 for the results.

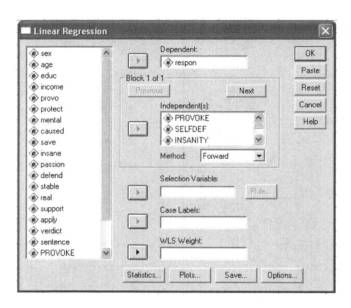

14.6.2   SPSS Syntax Method

COMPUTE PROVOKE=MEAN(PROVO,PASSION).
COMPUTE SELFDEF=MEAN(PROTECT,SAVE,DEFEND).
COMPUTE INSANITY=MEAN(MENTAL,INSANE,STABLE).
REGRESSION VARIABLES=(COLLECT)
/STATISTICS=DEFAULTS CHA TOL CI COLLIN
/DEPENDENT=RESPON
/FORWARD PROVOKE SELFDEF INSANITY.

14.6.3   SPSS Output

TABLE 14.1

Multiple Regression Analysis Output

Model	Variables Entered	Variables Removed	Method
		Variables Entered/Removed[a]	
1	SELFDEF		Forward (criterion: probability-of-F-to-enter <= .050)
2	INSANITY		Forward (criterion: probability-of-F-to-enter <= .050)
3	PROVOKE		Forward (criterion: probability-of-F-to-enter <= .050)

[a]Dependent variable: RESPON.

Model Summary

Model	R	R Square	Adjusted R Square	Std. Error of the Estimate	R Square Change	F Change	df1	df2	Sig. F Change
					Change Statistics				
1	.417[a]	174	.172	1.68	.174	81.920	1	389	.000
2	.431[b]	.186	.182	1.67	.012	5.740	1	388	.017
3	.446[c]	.198	.192	1.66	.012	6.032	1	387	.014

[a]Predictors: (constant), SELFDEF.
[b]Predictors: (constant), SELFDEF, INSANITY.
[c]Predictors: (constant), SELFDEF. INSANITY, PROVOKE.

ANOVA[d]

Model		Sum of Squares	df	Mean Square	F	Sig.
1	Regression	230.785	1	230.785	81.920	.000[a]
	Residual	1095.891	389	2.817		
	Total	1326.675	390			
2	Regression	246.761	2	123.380	44.329	.000[b]
	Residual	1079.915	388	2.783		
	Total	1326.675	390			
3	Regression	263.334	3	87.778	31.946	.000[c]
	Residual	1063.342	387	2.748		
	Total	1326.675	390			

[a]Predictors: (constant), SELFDEF.
[b]Predictors: (constant), SELFDEF, INSANITY.
[c]Predictors: (constant), SELFDEF, INSANITY, PROVOKE.
[d]Dependent variable: RESPON.

Coefficients[a]

Model		B	Std. Error	Beta	t	Sig.	Lower Bound	Upper Bound	Tolerance	VIF
		Unstandardized Coefficients		Standardized Coefficients			95% Confidence Interval for B		Collinearity Statistics	
1	(Constant)	5.970	.267		22.394	.000	5.446	6.495		
	SELFDEF	−.443	.049	−.417	−9.051	.000	−.539	−.346	1.000	1.000
2	(Constant)	5.441	.345		15.769	.000	4.763	6.119		
	SELFDEF	−.431	.049	−.406	-8.813	.000	−.527	−.335	.990	1.011
	INSANITY	.129	.054	.110	2.396	.017	.023	.234	.990	1.011
3	(Constant)	5.905	.391		15.086	.000	5.135	6.674		
	SELFDEF	−.408	.049	−.385	−8.262	.000	−.505	−.311	.956	1.046
	INSANITY	.155	.054	.132	2.839	.005	.048	.262	.953	1.050
	PROVOKE	−.129	.053	−.115	−2.456	.014	−.232	−.026	.937	1.067

[a]Dependent variable: respon.

Excluded Variables[c]

Model		Beta In	t	Sig.	Partial Correlation	Tolerance	VIF	Minimum Tolerance
						Collinearity Statistics		
1	PROVOKE	−.090[a]	−1.928	.055	−.097	.973	1.028	.973
	INSANITY	.110[a]	2.396	.017	.121	.990	1.011	.990
2	PROVOKE	−.115[b]	−2.456	.014	−.124	.937	1.067	.937

[a]Predictors in the model: (constant), SELFDEF.
[b]Predictors in the model: (constant), SELFDEF, INSANITY.
[c]Dependent variable: respon.

Collinearity Diagnostics[a]

Model	Dimension	Eigenvalue	Condition Index	(Constant)	SELFDEF	INSANITY	PROVOKE
				Variance Proportions			
1	1	1.948	1.000	.03	.03		
	2	.052	6.118	.97	.97		
2	1	2.815	1.000	.01	.01	.02	
	2	.146	4.384	.01	.27	.63	
	3	.039	8.534	.99	.72	.35	
3	1	3.746	1.000	.00	.01	.01	.01
	2	.148	5.032	.00	.21	.67	.01
	3	.073	7.185	.00	.39	.15	.75
	4	.034	10.545	.99	.39	.18	.24

[a]Dependent variable: respon.

14.6.4   Results and Interpretation

14.6.4.1   *Prediction Equation (Predicting the Level of Responsibility from the Three Defense Strategies of PROVOKE, SELFDEF, and INSANITY)*

The prediction equation is:

$$Y' = A + B_1X_1 + B_2X_2 + \ldots B_nX_n$$

where Y' = the predicted dependent variable, A = constant, B = unstandardized regression coefficient, and X=value of the predictor variable.

The relevant information for calculating the predicted responsibility attribution is presented in the **Coefficients** table (see Table 14.1). An examination of this table shows that all three predictor variables were entered into the prediction equation (Model 3), indicating that the defense strategies of self-defense, insanity, and provocation are significant predictors of the level of responsibility attributed to the battered woman for her fatal action. To predict the level of responsibility attributed from these three defense strategies, use the values presented in the **Unstandardized Coefficients** column for Model 3. Using the **Constant** and **B** (unstandardized coefficient) values, the prediction equation would be:

$$\text{Predicted responsibility attribution} = 5.91 + (-0.41 \times \text{SELFDEF}) + (0.16 \times \text{INSANITY}) + (-0.13 \times \text{PROVOKE})$$

Thus, for a subject who strongly believes that the battered woman's action was motivated by self-defense and provocation (a score of 8 on both scales) and not by an impaired mental state (a score of 1 on the insanity scale), the predicted level of responsibility attribution would be:

$$\text{Predicted responsibility attribution} = 5.91 + (-0.41 \times 8) + (0.16 \times 1) + (-0.13 \times 8)$$

$$= 5.91 - 3.28 + 0.16 - 1.04$$

$$= 1.75$$

Given that responsibility attribution is measured on an 8-point scale with 1 = not at all responsible to 8 = entirely responsible, a predicted value of 1.75 would suggest that this subject would attribute little responsibility to the battered woman for her fatal action.

14.6.4.2   *Evaluating the Strength of the Prediction Equation*

A measure of the strength of the computed prediction equation is **R-square**, sometimes called the **coefficient of determination**. In the regression model, R-square is the square of the correlation coefficient between Y, the observed value of the dependent variable, and Y', the predicted value of Y from the fitted regression line. Thus, if for each subject, the researcher computes the predicted responsibility attribution, and then calculates the square of the correlation coefficient between predicted responsibility attribution and observed responsibility attribution values, R-square is obtained. If all the observations fall on the regression line, R-square is 1 (perfect linear relationship). An R-square of 0 indicates no linear relationship between the predictor and dependent variables. To test the hypothesis of no linear relationship between the predictor and dependent variables, i.e., R-square = 0, the **Analysis of Variance** (ANOVA) is used. In this example, the results from this test are presented in the **ANOVA** table (see Table 14.1). The F value serves to test how well the regression model (Model 3) fits the data. If the probability associated with the F statistics is small, the hypothesis that R-square = 0 is rejected. For this example, the computed F statistic is 31.95, with an observed significance level of less than 0.001. Thus, the hypothesis that there is no linear relationship between the predictor and dependent variables is rejected.

14.6.4.3   *Identifying Multicollinearity*

When the predictor variables are correlated among themselves, the unique contribution of each predictor variable is difficult to assess. This is because of the overlapped or shared variance between the predictor variables, i.e., they are multicollinear. For this example, both the "tolerance" values (greater than 0.10) and the "VIF" values (less than 10) are all quite acceptable (see **Coefficients** table). Thus, multicollinearity does not seem to be a problem for this example.

Another way of assessing if there is too much multicollinearity in the model is to look at the **Collinearity Diagnostics** table. The *condition index* summarizes the findings, and a common rule of thumb is that a condition index of over 15 indicates a possible multicollinearity problem and a condition index of over 30 suggests a serious multicollinearity problem. For this example, multicollinearity is not a problem.

14.6.4.4   *Identifying Independent Relationships*

Once it has been established that multicollinearity is not a problem, multiple regression can be used to assess the relative contribution (independent relationship) of each predictor variable by *controlling the effects of other predictor variables* in the prediction equation. The procedure can also be used to assess the size and direction of the obtained independent relationships.

In identifying independent relationships, it is inappropriate to interpret the unstandardized regression coefficients (**B** values in the **Coefficients** table) as indicators of the relative importance of the predictor variables. This is because the **B** values are based on the actual units of measurement, which may differ from variable to variable, i.e., one variable may be measured on a five-point scale, whereas another variable may be measured on an eight-point scale. When variables differ substantially in units of measurement, the sheer magnitude of their coefficients does not reveal anything about their relative importance. Only if all predictor variables are measured in the same units are their coefficients directly comparable. One way to make regression coefficients somewhat more comparable is to calculate **Beta** weights, which are the coefficients of the predictor variables when all variables are expressed in standardized form (z-score).

In this example, the **Beta** weights (standardized regression coefficients) for all three defense strategies are presented in the **Coefficients** table. The size of the Beta weights indicates the strength of their independent relationships. From the table, it can be seen that the defense of self-defense has the strongest relationship with responsibility attribution, whereas the other two defense strategies — provocation and insanity — are weaker. The direction of the coefficients also shed light on the nature of the relationships. Thus, for the defense strategies of self-defense and provocation, the negative coefficients indicate that the more the subjects interpreted the battered woman's motive for her fatal action as being due to self-defense and provocation, the less they held her responsible for her action (self-defense: Beta = -0.39, t = 8.26, $p < .001$; provocation: Beta = -0.12, t = -2.46, $p < .05$). Conversely, the positive coefficient associated with the insanity variable shows that the more they interpreted the battered woman's action as being due to an impaired mental state, the more they held the woman responsible for her fatal action? (Beta = 0.13, t = 2.84, $p < .01$).

14.7 Example 2 — Hierarchical Regression

Another way of assessing the relative importance of predictor variables is to consider the increase in R-square when a variable is entered into an equation that already contains the other predictor variables. The increase in R-square resulting from the entry of a subsequent predictor variable indicates the amount of unique information in the dependent variable that is accounted for by that variable, *above and beyond* what has already been accounted for by the other predictor variables in the equation.

Suppose that the researcher is interested in comparing the relative importance of two sets of variables in predicting responsibility attribution to the battered woman in the previous example. Specifically, the researcher wants to assess the relative importance of a set of variables comprising the three defense strategies (self-defense, provocation, and insanity) and a set of variables comprising the subjects' demographic characteristics (sex, age, educational level, and income) in predicting responsibility attribution. This task can be accomplished by the use of **hierarchical regression** procedure. For this model, the researcher determines the order of entry for the two sets of predictor variables. Assume that the researcher believes that the subjects' demographics would be less strongly related to the dependent variable than the set of defense strategies. On the basis of this assumption, the researcher accords priority of entry into the prediction equation to the set of demographic variables, followed by the set of defense strategies. This order of entry will assess (1) the importance of the demographic variables in predicting the dependent variable of responsibility attribution, and (2) the amount of unique information in the dependent variable that is accounted for by the three defense strategies. This example employs the data set **DOMES.SAV**.

14.7.1   Windows Method

1. From the menu bar, click **Analyze**, then **Regression**, and then **Linear…**. The following **Linear Regression** window will open.

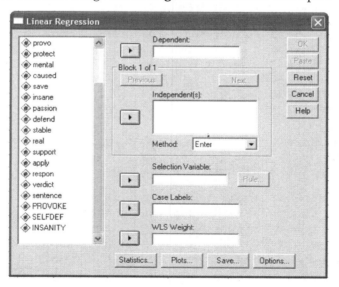

2. Click (highlight) the variable of **RESPON** and then click 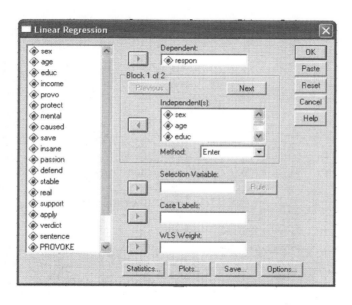 to transfer this variable to the **Dependent** field. Because the set of demographic variables (**SEX, AGE, EDUCATIONAL LEVEL,** and **INCOME**) will be entered first into the prediction equation (**Block 1**), click (highlight) these variables and then click to transfer these variables to the **Independent(s)** field. In the **Method** cell, select **ENTER** from the drop-down list as the method of entry for this set of demographic variables into the prediction equation.

Next, click to open **Block 2** in the **Independent(s)** field for entry of the second set of independent variables. Click (highlight) the variables of **PROVOKE, SELFDEF,** and **INSANITY,** and then click to transfer these variables to the **Independent(s)** field. In the **Method** field, select **ENTER** from the drop-down list as the method of entry for this set of independent (predictor) variables into the prediction equation.

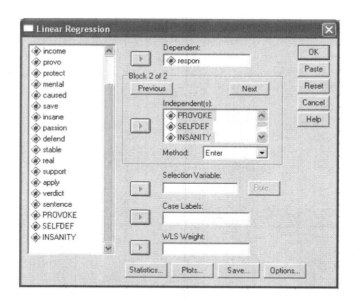

3. Click 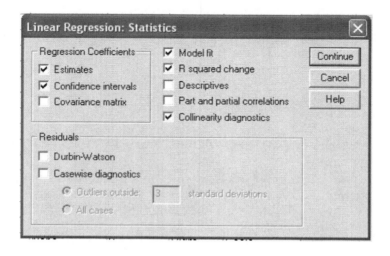 to open the **Linear Regression: Statistics** window. Check the fields to obtain the statistics required. For this example, check the fields for **Estimates, Confidence intervals, Model fit, R squared change**, and **Collinearity diagnostics**. Click ▭Continue▭ when finished.

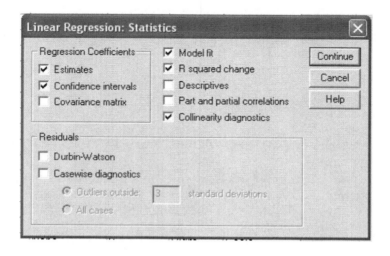

4. When the **Linear Regression** window opens, click 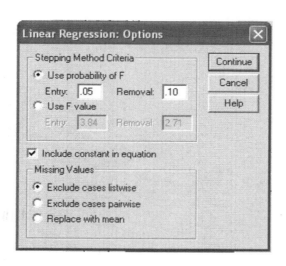 to open the **Linear Regression: Options** window. Ensure that both the **Use probability of F** and the **Include constant in equation** fields are checked. Under **Missing Values**, check the **Exclude cases listwise** field. Click to return to the **Linear Regression** window.

5. When the **Linear Regression** window opens, click OK to complete the analysis. See Table 14.2 for the results.

14.7.2 SPSS Syntax Method

REGRESSION VARIABLES=(COLLECT)
/STATISTICS=DEFAULTS CHA TOL CI COLLIN
/DEPENDENT=RESPON
/METHOD=ENTER SEX AGE EDUC INCOME/ENTER PROVOKE
SELFDEF INSANITY.

14.7.3   SPSS Output

TABLE 14.2

Hierarchical Multiple Regression Output

Variables Entered/Removed[b]

Model	Variables Entered	Variables Removed	Method
1	INCOME, EDUC, SEX, AGE[a]	.	Enter
2	PROVOKE, INSANIT SELFDEF[a]	.	Enter

[a]All requested variables entered.
[b]Dependent variable: RESPON.

Model Summary

Model	R	R Square	Adjusted R Square	Std. Error of the Estimate	Change Statistics R Square Change	F Change	df1	df2	Sig. F Change
1	.241[a]	.058	.048	1.80	.058	5.727	4	372	.000
2	.469[b]	.220	.205	1.65	.162	25.481	3	369	.000

[a]Predictors: (constant). INCOME, EDUC, SEX, AGE.
[b]Predictors: (constant), INCOME, EDUC, SEX, AGE, PROVOKE, INSANITY, SELFDEF.

ANOVA[c]

Model		Sum of Squares	df	Mean Square	F	Sig.
1	Regression	74.280	4	18.570	5.727	.000[a]
	Residual	1206.261	372	3.243		
	Total	1280.541	376			
2	Regression	281.286	7	40.184	14.839	.000[b]
	Residual	999.255	369	2.708		
	Total	1280.541	376			

[a]Predictors: (constant), INCOME, EDUC, SEX, AGE.
[b]Predictors: (constant), INCOME, EDUC, SEX, AGE, PROVOKE, INSANITY, SELFDEF.
[c]Dependent variable: RESPON.

Coefficients[a]

Model		Unstandardized Coefficients		Standardized Coefficients	t	Sig.	95% Confidence Interval for B		Collinearity Statistics	
		B	Std. Error	Beta			Lower Bound	Upper Bound	Tolerance	VIF
1	(Constant)	4.458	.658		6.770	.000	3.163	5.753		
	sex	-.763	.200	-.192	-3.811	.000	-1.157	-.369	.994	1.006
	ae	-.019	.013	-.086	-1.511	.132	-.044	.006	.784	1.276
	educ	.126	.086	.076	1.470	.142	-.042	.294	.958	1.044
	income	.221	.139	.089	1.590	.113	-.052	.494	.802	1.246
2	(Constant)	6.633	.719		9.220	.000	5.219	8.048		
	sex	-.521	.185	-.131	-2.810	.005	-.885	-.156	.969	1.032
	age	-.019	.012	-.086	-1.652	.099	-.042	.004	.783	1.277
	educ	.069	.078	.041	.876	.381	-.086	.223	.950	1.053
	income	.107	.128	.043	.835	.404	-.145	.359	.791	1.265
	PROVOKE	-.121	.053	-.109	-2.295	.022	-.224	-.017	.931	1.074
	SELFDEF	-.382	.050	-.361	-7.591	.000	-.481	-.283	.936	1.068
	INSANITY	.127	.056	.108	2.290	.023	.018	.236	.943	1.061

[a]Dependent variable: respon.

Excluded Variables

Model		Beta In	F	Sig.	Partial Correlation	Collinearity Statistics Tolerance
1	PROVOKE	−.145[a]	8.434	.004	−.149	.992
	SELFDEF	−.385[a]	66.421	.000	−.390	.967
	INSANITY	.110[a]	4.705	.031	.112	.983

[a]Predictors in the model: (constant), INCOME, EDUC, SEX, AGE.
[b]Dependent variable: RESPON.

Collinearity Diagnostics[a]

Model	Dimension	Eigenvalue	Condition Index	(Constant)	sex	age	educ	Income	PROVOKE	SELFDEF	INSANITY
1	1	4.634	1.000	.00	.00	.00	.00	.01			
	2	.208	4.722	.01	.05	.01	.02	.65			
	3	.084	7.429	.00	.00	.75	.09	.26			
	4	.060	8.786	.01	.70	.05	.23	.08			
	5	.014	18.109	.98	.25	.18	.66	.00			
2	1	7.284	1.000	.00	.00	.00	.00	.00	.00	.00	.00
	2	.252	5.380	.00	.01	.03	.00	.55	.02	.03	.01
	3	.157	6.805	.00	.03	.02	.00	.00	.00	.09	.72
	4	.084	9.296	.00	.03	.63	.01	.39	.05	.16	.01
	5	.080	9.542	.00	.20	.12	.12	.00	.37	.08	.00
	6	.074	9.908	.00	.01	.01	.03	.00	.46	.57	.18
	7	.057	11.304	.01	.56	.04	.32	.06	.02	.00	.05
	8	.011	25.521	.99	.16	.15	.52	.00	.08	.08	.04

Variance Proportions span over: (Constant) sex age educ Income PROVOKE SELFDEF INSANITY

[a]Dependent variable: respon.

14.7.4 Results and Interpretation

In the **Model Summary** table (see Table 14.2), Model 1 represents entry of the first set of demographic variables, and Model 2 represents entry of the second set of self-defense strategy variables. The results show that Model 1 (demographics) accounted for 5.8% of the variance (**R Square**) in the subjects' responsibility attribution. Entry of the three defense strategy variables (Model 2) resulted in an **R Square Change** of 0.162. This means that entry of the three defense strategy variables increased the explained variance in the subjects' responsibility attribution by 16.2% to a total of 22%. This increase is significant by the **F Change** test, $F(3,369) = 25.48$, $p < .001$ (a test for the increase in explanatory power). These results suggest that the defense strategy variables represent a significantly more powerful set of predictors than the set of demographic variables.

In the **ANOVA** table, the results show that entry of the set of demographic variables alone (Model 1) yielded a significant prediction equation, $F(4,372) = 5.73$, $p < .001$. Addition of the three defense strategy variables (Model 2) resulted in an overall significant prediction equation, $F(7,369) = 14.84$, $p < .001$.

Looking at Model 2 in the **Coefficients** table, it can be seen that multicollinearity is not a problem — all tolerance values are above 0.10; all VIF values are below 10; and the condition indices for the seven predictor variables are below 15. In examining the **Beta** weights (standardized regression coefficients), it can also be seen that all three defense strategy variables are significant predictors of responsibility attribution ($p < .05$), whereas subjects' sex is the only demographic variable that was found to be significant. Thus, the more the subjects perceived the battered woman's fatal action was due to provocation ($\beta = -0.11$, $t = -2.30$, $p < 0.05$), and to self-defense ($\beta = -0.36$, $t = -7.59$, $p < .001$), the less responsibility they attributed to the woman for her action. Conversely, the more the subjects perceived the battered woman's fatal action was due to insanity, the greater the responsibility they attributed to the woman for her action ($\beta = 0.11$, $t = 2.29$, $p < .05$). The finding that subjects' sex was a significant predictor ($\beta = -0.13$, $t = -2.81$, $p < .01$), indicated that males attributed greater responsibility to the woman for her fatal action than females did (code: 1 = male, 2 = female).

14.8   Example 3 — Path Analysis

With path analysis, multiple regression is often used in conjunction with a causal theory, with the aim of describing the entire structure of linkages between independent and dependent variables posited from that theory. For example, based on theoretical considerations of the domestic violence example, a researcher has constructed the path model presented in Figure 14.1 to represent the hypothesized structural relationships between the three defense strategies of provocation, self-defense, and insanity, and the attribution of responsibility.

The model specifies an "ordering" among the variables that reflects a hypothesized structure of cause-effect linkages. Multiple regression techniques can be used to determine the magnitude of *direct* and *indirect* influences that each variable has on other variables that follow it in the presumed

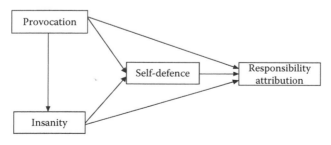

FIGURE 14.1

causal order (as indicated by the directional arrows). Each arrow in the model represents a presumed causal linkage or path of causal influence. Through regression techniques, the strength of each separate path can be estimated. This analysis actually involves three regression equations because (1) responsibility attribution is a dependent variable for the three defense variables, (2) the self-defense variable is a dependent variable for the defense variables of provocation and insanity, and (3) the insanity defense variable is a dependent variable for the provocation defense variable. This example employs the data set, **DOMES.SAV**.

14.8.1   Windows Method

1. The first regression equation involves predicting the level of responsibility (**RESPON**) attributed to the battered woman's action from all three defense strategies (**PROVOKE, SELFDEF**, and **INSANITY**). Click **Analyze**, then **Regression**, and then **Linear...** from the menu bar. The following **Linear Regression** window will open.

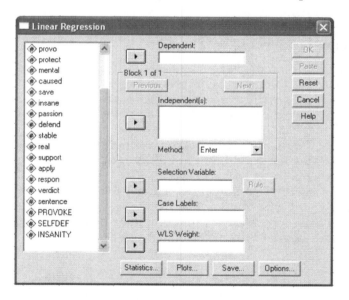

2. Click (highlight) the variable of **RESPON**, and then click 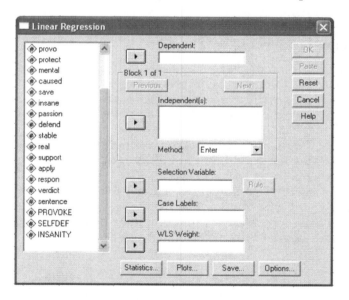 to transfer this variable to the **Dependent** field. Next, click (highlight) the three variables of **PROVOKE, SELFDEF**, and **INSANITY**, and then click to transfer these variables to the **Independent(s)** field. In the **Method** field, select **FORWARD** from the drop-down

list as the method of entry for the three independent (predictor) variables into the prediction equation.

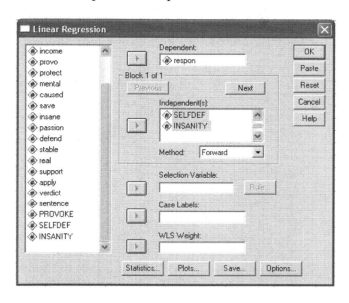

3. Click 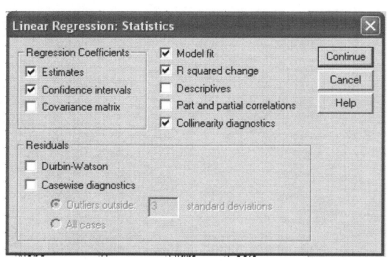 Statistics... to open the **Linear Regression: Statistics** window. Check the fields to obtain the statistics required. For this example, check the fields for **Estimates**, **Confidence intervals**, **Model fit**, **R squared change**, and **Collinearity diagnostics**. Click Continue when finished.

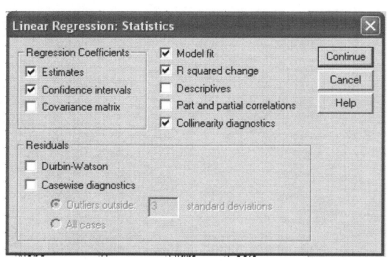

4. When the **Linear Regression** window opens, click Options... to open the **Linear Regression: Options** window. Ensure that the **Use probability of F**, the **Include constant in equation**, and **Exclude cases listwise** fields are checked.

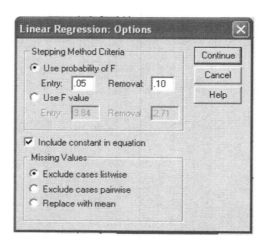

Click Continue to return to the **Linear Regression** window.

5. When the **Linear Regression** window opens, click OK to complete the analysis.

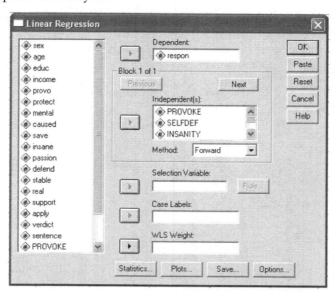

6. The second prediction equation involves predicting the defense strategy of **SELFDEF** from the defense strategies of **PROVOKE** and **INSANITY**. To do this, repeat step 1 to step 5. In step 1, transfer **SELFDEF** to the **Dependent** field, and the variables of **PROVOKE**

and **INSANITY** to the **Independent(s)** field. Click ![OK] to complete the analysis.

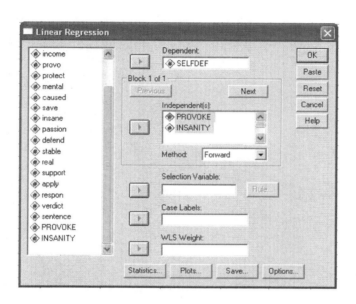

7. The third prediction equation involves predicting the defense strategy of **INSANITY** from the defense strategy of **PROVOKE**. To do this, repeat step 1 to step 5. In step 1, transfer **INSANITY** to the **Dependent** field, and the variable of **PROVOKE** to the **Independent(s)** field. Click ![OK] to complete the analysis. See Table 14.3 for the results.

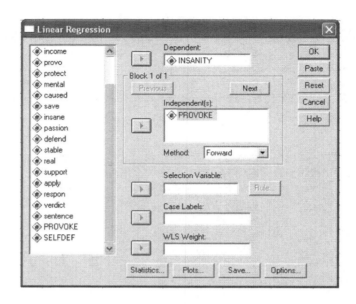

14.8.2   SPSS Syntax Method

REGRESSION VARIABLES=(COLLECT)
/STATISTICS=DEFAULTS CHA TOL CI COLLIN
/DEPENDENT=RESPON
/FORWARD PROVOKE SELFDEF INSANITY.
REGRESSION VARIABLES=(COLLECT)
/STATISTICS=DEFAULTS CHA TOL CI COLLIN
/DEPENDENT=SELFDEF
/FORWARD PROVOKE INSANITY.
REGRESSION VARIABLES=(COLLECT)
/STATISTICS=DEFAULTS CHA TOL CI COLLIN
/DEPENDENT=INSANITY
/FORWARD PROVOKE.

14.8.3 SPSS Output

TABLE 14.3

Path Analysis Output

Regression

Model	Variables Entered	Variables Removed	Method
	Variables Entered/Removed[a]		
1	SELFDEF		Forward (criterion: probability-of-F-to-enter <= .050)
2	INSANITY		Forward (criterion: probability-of-F-to-enter <= .050)
3	PROVOKE		Forward (criterion: probability-of-F-to-enter <= .050)

[a]Dependent variable: RESPON.

Model Summary

Model	R	R Square	Adjusted R Square	Std. Error of the Estimate	R Square Change	F Change	df1	df2	Sig. F Change
					Change Statistics				
1	.417[a]	.174	.172	1.68	.174	81.920	1	389	.000
2	.431[b]	.186	.182	1.67	.012	5.740	1	388	.017
3	.446[c]	.198	.192	1.66	.012	6.032	1	387'	.014

[a]Predictors: (constant), SELFDEF.
[b]Predictors: (constant), SELFDEF, INSANITY.
[c]Predictors: (constant), SELFDEF, INSANITY, PROVOKE.

ANOVA[d]

Model		Sum of Squares	df	Mean Square	F	Sig.
1	Regression	230.785	1	230.785	81.920	.000[a]
	Residual	1095.891	389	2.817		
	Total	1326.675	390			
2	Regression	246.761	2	123.380	44.329	.000[b]
	Residual	1079.915	388	2.783		
	Total	1326.675	390			
3	Regression	263.334	3	87.778	31.946	.000[c]
	Residual	1063.342	387	2.748		
	Total	1326.675	390			

[a]Predictors: (constant), SELFDEF.
[b]Predictors: (constant), SELFDEF, INSANITY.
[c]Predictors: (constant), SELFDEF, INSANITY, PROVOKE.
[d]Dependent variable: RESPON.

Coefficients[a]

Model		Unstandardized Coefficients B	Std. Error	Standardized Coefficients Beta	t	Sip.	95% Confidence Interval for B Lower Bound	Upper Bound	Collinearity Statistics Toler-ance	VIF
1	(Constant)	5.970	.267		22.394	.000	5.446	6.495		
	SELFDEF	−.443	.049	−.417	−9.061	.000	−.539	−.346	1.000	1.000
2	(Constant)	5.441	.345		15.769	.000	4.763	6.119		
	SELFDEF	−.431	.049	−.406	−8.813	.000	−.527	−.335	.990	1.011
	INSANITY	.129	.054	.110	2.396	.017	.023	.234	.990	1.011
3	(Constant)	5.905	.391		15.086	.000	5.135	6.674		
	SELFDEF	−.408	.049	−.385	−8.262	.000	−.505	−.311	.956	1.046
	INSANITY	.155	.054	.132	2.839	.005	.048	.262	.953	1.050
	PROVOKE	−.129	.053	−.115	−2.456	.014	−.232	−.026	.937	1.067

[a]Dependent variable: respon.

Excluded Variable[c]

Model		Beta In	F	Sig.	Partial Correlation	Collinearity Statistics Tolerance
1	PROVOKE	−.090[a]	3.716	.055	−.097	.973
	INSANITY	.110[a]	5.740	.017	.121	.990
2	PROVOKE	−.115[b]	6.032	.014^	−124	.937[7]

[a]Predictors in the model: (constant), SELFDEF.
[b]Predictors in the model: (constant), SELFDEF, INSANITY.
[c]Dependent variable: RESPON.

Collinearity Diagnostics[a]

Model	Dimension	Eigen value	Condition Index	Variance Proportions (Constant)	SELF DEF	INSANITY	PROVOKE
1	1	1.948	1.000	.03	.03		
	2	.052	6.118	.97	.97		
2	1	2.815	1.000	.01	.01	.02	
	2	.146	4.384	.01	.27	.63	
	3	.039	8.534	.99	.72	.35	
3	1	3.746	1.000.	.00	.01	.01	.01
	2	.148	5.032	.00	.21	.67	.01
	3	.073	7.185	.00	.39	.15	.75
	4	.034	10.545	.99	.39	.18	.24

[a]Dependent variable: respon.

Regression

| | | Variables Entered/Removed[a] | |
Model	Variables Entered	Variables Removed	Method
1	PROVOKE		Forward (criterion: probability-of-F-to-enter <= .050)
2	INSANITY		Forward (criterion: probability-of-F-to-enter <= .050)

[a]Dependent variable: SELFDEF.

				Model Summary				
				Std.		Change Statistics			
				Error of	R				
		R	Adjusted	the	Square				Sig. F
Model	R	Square	R Square	Estimate	Change	F Change	df1	df2	Change
1	.158[a]	.025	.023	1.7204	.025	10.117	1	394	.002
2	.205[b]	.042	.037	1.7077	.017	6.896	1	393	.009

[a]Predictors: (constant), PROVOKE.
[b]Predictors: (constant), PROVOKE, INSANITY.

Model		Sum of Squares	df	Mean Square	F	Sig.
1	Regression	29.946	1	29.946	10.117	.002[a]
	Residual	1166.211	394	2.960		
	Total	1196.157	395			
2	Regression	50.057	2	25.028	8.582	.000[b]
	Residual	1146.101	393	2.916		
	Total	1196.157	395			

ANOVA[c]

[a]Predictors: (constant), PROVOKE.
[b]Predictors: (constant), PROVOKE, INSANITY.
[c]Dependent variable: SELFDEF.

Coefficients[a]

Model		Unstandardized Coefficients		Standardized Coefficients			95% Confidence Interval for B		Collinearity Statistics	
		B	Std. Error	Beta	t	Sip.	Lower Bound	Upper Bound	Tolerance	VIF
1	(Constant)	4.310	.286		15.084	.000	3.748	4.872		
	PROVOKE	.166	.052	.158	3.181	.002	.064	.269	1.000	1.000
2	(Constant)	4.715	.323		14.603	.000	4.081	5.350		
	PROVOKE	.190	.053	.180	3.602	.000	.086	.293	.972	1.029
	INSANITY	−.145	.055	−.132	−2.626	.009	−.254	−.036	.972	1.029

[a]Dependent variable: SELFDEF.

Excluded Variables[b]

Model		Beta In	F	Sig.	Partial Correlation	Collinearity Statistics Tolerance
1	INSANITY	−.132[a]	6.896	.009	−.131	.972

[a]Predictors in the model: (constant), PROVOKE.
[b]Dependent variable: SELFDEF.

Collinearity Diagnostics[a]

Model	Dimension	Eigenvalue	Condition Index	Variance Proportions		
				(Constant)	PROVOKE	INSANITY
1	1	1.953	1.000	.02	.02	
	2	.047	6.455	.98	.98	
2	1	2.844	1.000	.01	.01	.02
	2	.112	5.035	.04	.23	.89
	3	.044	8.022	.95	.76	.09

[a]Dependent variable: SELFDEF.

Regression

Variables Entered/Removed

Model	Variables Entered	Variables Removed	Method
1	PROVOKE		Forward (criterion: probability-of-F-to-enter <= .050)

aDependent variable: INSANITY.

Model Summary

Model	R	R Square	Adjusted R Square	Std. Error of the Estimate	R Square Change	F Change	df1	df2	Sig. F Change
1	.169a	.028	.026	1.5579	.028	11.546	1	394	.001

| | | | | | Change Statistics | | | | |

aPredictors: (constant), PROVOKE.

ANOVAb

Model		Sum of Squares	df			
1	Regression	28.020	1	28.020	11.546	.001a
	Residual	956.198	394	2.427		
	Total	984.218	395			

aPredictors: (constant), PROVOKE.
bDependent variable: INSANITY.

Coefficients[a]

Model		Unstandardized Coefficients		Standardized Coefficients			95% Confidence Interval for B		Collinearity Statistics	
		B	Std. Error	Beta	t	Sig.	Lower Bound	Upper Bound	Tolerance	VIF
1	(Constant)	2.795	.259		10.803	.000	2.286	3.304		
	PROVOKE	.161	.047	.169	3.398	.001	.068	.254	1.000	1.000

[a]Dependent variable: INSANITY

Collinearity Diagnostics[a]

Model	Dimension	Eigen value	Condition Index	Variance Proportions	
				(Constant)	PROVOKE
1	1	1.953	1.000	.02	.02
	2	.047	6.455	.98	.98

[a]Dependent variable: INSANITY.

14.8.4   Results and Interpretation

The path model depicted in Figure 14.1 hypothesizes that subjects' interpretation of the battered woman's motives as provocation will have both *direct* and *indirect* influences on their responsibility attribution; the indirect influence being mediated by their endorsement of the insanity and self-defense strategies. The direction of the arrows depicts the hypothesized direct and indirect paths. To estimate the magnitude of these paths, a series of regression analyses were carried out.

1. The path coefficients between responsibility attribution and the three defense strategies were obtained by regressing the former on the latter. The results from the **Coefficients** table (see Table 14.3) generated from the first regression analysis show that all three defense strategies entered the prediction equation (Model 3) (i.e., all three defense strategies are significant predictors). The Beta values presented in the **Standardized Coefficients** column represent the standardized regression coefficients between responsibility attribution and the three defense strategies (SELFDEF: Beta = –0.39; INSANITY: Beta = 0.13; PROVOKE: Beta = –0.12).

2. The path coefficients between the self-defense strategy and the other two defense strategies of provocation and insanity were obtained by regressing the former on the latter. The results from the **Coefficients** table generated from the second regression analysis show that both provocation and insanity defenses are significant predictors of the defense of self-defense (PROVOKE: Beta = 0.18; INSANITY: Beta = –0.13).

3. The path coefficient between the defense strategies of insanity and provocation is presented in the **Coefficients** table generated from the third regression analysis, and is significant (PROVOKE: Beta = 0.17).

Figure 14.2 presents the path model together with the estimated regression coefficients (Beta values) associated with the hypothesized paths.

It can be concluded that the defense strategies of provocation and insanity have direct influences on the subjects' responsibility attribution. The direction of the regression coefficients indicates that (1) the more the subjects endorsed the provocation defense, the less responsibility they attributed to the battered woman for her fatal action, and (2) the more they endorsed the insanity defense, the more they held the battered woman responsible. The results also show that at least part of these influences is indirect, being mediated by the subjects' endorsement of the self-defense strategy. Thus, the more the subjects endorsed the provocation defense, the more they believed that the woman acted in self-defense, and which in turn is associated with a lower level of responsibility attribution. Endorsing the provocation defense also led to an increased belief that the battered woman's action was due to an impaired mental state, which in turn decreased the willingness to accept the plea of self-defense. This, in turn, is associated with a higher level of responsibility attributed to her.

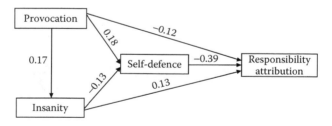

FIGURE 14.2

15

Structural Equation Modeling

15.1 What Is Structural Equation Modeling?

Structural equation modeling (SEM) is a multivariate technique that can best be described as a combination of both factor analysis and path analysis. It is a statistical technique that allows the analyst to examine a series of dependence relationships between **exogenous variables** and **endogenous variables** simultaneously. Before proceeding on to explaining the characteristics of SEM, it is necessary to comment on the distinction between exogenous and endogenous variables when used in causal models. According to Pedhazur (1997), an **exogenous** variable is one whose variability is assumed to be determined by causes outside the causal model under consideration. An **endogenous** variable, on the other hand, is one whose variation is to be explained by exogenous and other endogenous variables in the causal model.

The usefulness of SEM in research is distinguished by three characteristics:

1. **It provides a method of dealing with multiple relationships simultaneously.** As shown in Chapter 14, model testing via path analysis can be carried out with the conventional multiple regression technique. That is, path coefficients can be estimated by regressing the endogenous variable (the dependent variable) on to the exogenous variables (the predictor variables), and then repeating the procedure by treating the exogenous variables as endogenous variables. Such a method, however, is typically piecemeal in nature and does not provide information regarding the hypothesized model's goodness-of-fit. Without information about the model's goodness-of-fit, it is difficult to assess the adequacy of the theory underlying the hypothesized model. SEM, on the other hand, is able to estimate the multiple and interrelated dependence relationships simultaneously. Because it tests the model as a whole, rather than in a piecemeal fashion, statistics can be calculated to show the goodness-of-fit of the data to the hypothesized model.

2. **It is able to represent unobserved (latent) concepts in the analysis of dependence relationships.** Although multiple regression can be used to analyze relationships between variables, its use is limited to the analysis of those variables that can only be directly observed (or measured). SEM, on the other hand, has the ability to incorporate latent (or unobserved) variables in the analysis. A **latent** variable is a hypothesized or unobserved construct and as such, cannot be measured directly. It can only be approximated by observable or measured variables. For example, a researcher wants to investigate the pattern of relationships between three psychological constructs: *aggression, authoritarianism,* and *intelligence.* All three constructs cannot be directly observed and, therefore, cannot be directly measured. They are "measured" indirectly using various types of scale items, from questionnaires and inventories. Based on the responses to these scale items, the magnitude of these latent variables can be estimated. This is similar to factor analysis, in which highly interrelated clusters of scale items are identified as latent factors. SEM can then be used to estimate paths between latent factors rather than between variables.

3. **It improves statistical estimation by accounting for measurement error in the estimation process.** The univariate and multivariate techniques covered in the previous chapters all assume that there is no error associated with the measurement of variables. That is, these techniques assume that variables in the analyses are error-free. Although this is an unrealistic assumption, many if not most researchers, regrettably, act as if this is indeed the case for their data. We know from both theoretical and practical perspectives that concepts can seldom be measured perfectly, either because of inaccurate responses by the respondents, or because of problems associated with the operationalization of the concepts. Consequently, measured variables usually contain at least moderate amounts of error, and when such measures are used in univariate and multivariate models (e.g., ANOVA, ANCOVA, MANOVA, and multiple regression), the coefficients obtained will be biased, most often in unknown degree and direction. SEM, on the other hand, uses scores on the measured or manifest variables to develop estimates of the individual's scores on the underlying construct, or latent variables. As these estimates are derived on the basis of the common or shared variance among the measured variables, scores on the latent variables are unaffected by random measurement error. Thus, when these variables are used in SEM analysis, the potentially biasing effects of random measurement error on the results are removed. The net result is that the statistical estimation process is improved, because the structural paths between latent variables are relatively free of the unreliabilities of their measurement indicators.

15.2   The Role of Theory in SEM

SEM is theory driven — that is, when a researcher employs SEM to test a theory, there is a strong need for justification for the specification of the dependence relationships. Theory provides this justification. For example, a researcher specifies the following model to represent the hypothesized relationships between the three variables of *authoritarianism*, *racism*, and *aggression*.

Authoritarianism → Racism → Aggression

For this model, the unidirectional arrows linking the three variables suggest a "causal" flow of influence, with authoritarianism hypothesized to influence the level of racism, and the level of racism hypothesized to influence the level of aggression. When specifying the pattern of relationships between the three variables, specification of the directional relationships (paths) must be guided by theoretical considerations; i.e., the specification does not arrive out of thin air. Without theoretical guidance, the fit of the hypothesized model will at best capitalize on chance, and at worse, is nonsensical.

The need for theory to guide the specification process becomes especially critical when model modifications are made. Because of the flexibility of SEM, the researcher can set many constraints on the model so as to make it "fit" better. But without theoretical guidance, the researcher may "overfit" the model by setting too many constraints. The result may be a model that fits well, but one which is too restricted and with little generalizability. From a practical perspective, a theory-based approach to SEM is a distinct strength of this technique, as it entails a mode of thinking that forces the researcher to specify the theoretical model employed more exactly, testing the theory more precisely, and yielding a more thorough understanding of the data.

15.3   The Structural Equation Model

In its most general form, SEM consists of two parts: the **measurement model** and the **structural equation model**.

The **measurement model** specifies the rules governing how the latent variables are measured in terms of the observed variables, and it describes the measurement properties of the observed variables. That is, measurement models are concerned with the relations between observed and latent variables. Such models specify hypotheses about the relations between a set of observed variables, such as ratings or questionnaire items, and the unobserved variables or constructs they were designed to measure.

The **structural equation model** is a flexible, comprehensive model that specifies the pattern of relationships among independent and dependent variables, either observed or latent. It incorporates the strengths of multiple regression analysis, factor analysis, and multivariate ANOVA (MANOVA) in a single model that can be evaluated statistically. Moreover, it permits directional predictions among a set of independent or a set of dependent variables, and it permits modeling of indirect effects.

Of the two models, the structural model is of greater interest to the researcher, because it offers a direct test of the theory of interest. The measurement model is important as it provides a test for the reliability of the observed variables employed to measure the latent variables. A measurement model that offers a poor fit to the data suggests that at least some of the observed indicator variables are unreliable, and precludes the researcher from moving to the analysis of the structural model.

15.4   Goodness-of-Fit Criteria

A number of goodness-of-fit measures are available to assess the overall fit of the hypothesized model. Goodness-of-fit measures the extent to which the actual or observed covariance input matrix corresponds with (or departs from) that predicted from the proposed model. Goodness-of-fit measures can be classified into three types: (1) absolute fit measures, (2) incremental fit measures, and (3) parsimonious fit measures. Examples of these measures are presented in the following subsections. These examples have been chosen not because they represent the "best" indicators of goodness-of-fit, but because they are probably the easiest to understand.

15.4.1   Absolute Fit Measures

These measures determine the degree to which the proposed model predicts (fits) the observed covariance matrix. Some commonly used measures of absolute fit include the **chi-square statistic**, the **goodness-of-fit statistic**, and the **root mean square error of approximation**.

- **Chi-square statistic:** The most fundamental measure of overall fit is the likelihood ratio chi-square (χ^2) statistic, the only statistically based measure of goodness-of-fit available in SEM (Jöreskog & Sörbom, 1993). In applying the chi-square test, the researcher customarily wishes to reject the null hypothesis so as to claim support for its alternative, i.e., there is a significant difference between the "observed" and the "expected." When applied in this way, the larger the chi-square value, the "better." However, when used in SEM, the researcher is looking for *insignificant differences* between the actual and predicted matrices. As such, the researcher does *not* wish to reject the null hypothesis and, accordingly, the smaller the chi-square value, the better fit of the model. However, the chi-square statistic is very sensitive to departures from multivariate normality of the observed variables and increases as a direct function of sample size. In case of large samples, the power of the statistical test underlying the SEM approach is very high. With a great deal of statistical power, almost every reasonable model will be rejected if only the chi-square value and the associated probability are considered. Therefore, given departures from multivariate normality or large samples, a proposed model can easily fail to fit the data statistically, even though the discrepancy between the sample covariance matrix and that reproduced by the parameter estimates of the proposed model may be insignificant from a practical point of view. Given these limitations, the researcher should complement the chi-square measure with other goodness-of-fit measures.

- **Goodness-of-Fit Index (GFI):** The GFI measures how much *better* the model fits compared with no model at all (Jöreskog & Sörbom, 1989). It is a nonstatistical measure ranging from 0 (poor fit) to 1 (perfect fit). Although higher values indicate a better fit, no threshold levels for acceptability have been established.

- **Root Mean Square Error of Approximation (RMSEA):** The RMSEA takes into account the error of approximation in the population. It is a measure of *discrepancy per degree of freedom*, and asks the question, "How well would the model, with unknown but optimally chosen values, fit the population covariance matrix if it were available?" (Browne & Cudeck, 1993, pp. 137–138). The value is representative of the goodness-of-fit when the proposed model is estimated in the population. Values ranging from 0.05 to 0.08 are deemed acceptable, values ranging from 0.08 to 0.10 indicate mediocre fit, and those greater than 0.10 indicate poor fit (Browne & Cudeck, 1993; MacCallum, Browne, & Sugawara, 1996).

15.4.2 Incremental Fit Measures

These measures compare the proposed model to some baseline model, most often referred to as the *null* or *independence* model. In the independence model, the observed variables are assumed to be uncorrelated with each other. The independence model is so severely and implausibly constrained that it would provide a poor fit to any interesting set of data. A number of incremental fit measures have been proposed, such as **Tucker-Lewis Index (TLI)**, **Normed Fit Index (NFI)**, **Relative Fit Index (RFI)**, **Incremental Fit Index (IFI)**, and **Comparative Fit Index (CFI)**. Although the calculations of these fit indices and their underlying assumptions may be somewhat different, they all represent comparisons between the proposed model and a null or independence model. Specifically, they show the improvement achieved by a proposed model over the null model (i.e., a model assuming independence among the variables); they range from 0 (a fit that is no better than the null model) to 1 (a perfect fit).

15.4.3   Parsimonious Fit Measures

In scientific research, theories should be as simple, or parsimonious, as possible. As Bentler and Moorjaart (1989) put it, "models with fewer unknown parameters may be considered as standing a better chance of being scientifically replicable and explainable" (page 315). Following this line of thought, parsimonious fit measures relate the goodness-of-fit of the proposed model to the number of estimated coefficients required to achieve this level of fit. Their basic objective is to diagnose whether model fit has been achieved by "overfitting" the data with too many coefficients. Their use is primarily to compare models on the basis of some criteria that take parsimony (in the sense of number of parameters to be estimated) as well as fit into account.

- **Parsimonious Normed Fit Index (PNFI):** The PNFI takes into account the number of degrees of freedom used to achieve a level of fit. Parsimony is defined as achieving higher degrees of fit per degree of freedom used (one degree of freedom per estimated coefficient). Higher values of PNFI are better, and its principal use is for the comparison of models with differing degrees of freedom. When comparing between models, differences of 0.06 to 0.09 are proposed to be indicative of substantial model differences (Williams & Holahan, 1994).

- **Akaike Information Criterion (AIC):** The AIC is a comparative measure between models with differing numbers of constructs. AIC values closer to zero indicate better fit and greater parsimony. A small AIC generally occurs when small chi-square values are achieved with fewer estimated coefficients. This shows not only a good fit of observed vs. predicted covariances, but also a model not prone to "overfitting." In applying this measure to the comparison decision problem, one estimates all models, ranks them according to the AIC criterion, and chooses the model with the smallest value.

15.4.4   Note of Caution in the Use of Incremental Fit Indices as "Rules of Thumb"

By convention, researchers have used incremental fit indices > 0.90 as traditional cutoff values to indicate acceptable levels of model fit. This is based on the logic that if a posited model achieves incremental fit indices > 0.90, then the model represents more than 90% improvement over the null or independence model. In other words, the only possible improvement to the model is less than 10%. The popularity of these indices as tools to assess the fit of models in covariance structure analyses lies with their ability to provide such absolute cutoff values that allow researchers to decide whether or not a model adequately fits the data, which have broad generality across different conditions and sample sizes.

Recently, Marsh and his colleagues (Marsh, Hau, & Wen, 2004; McDonald and Marsh, 1990) have sounded a note of warning about relying on these traditional cutoff values as "rules of thumb" to assess model fit across different research conditions and sample sizes. Their warning is consistent with Hu and Bentler's (1998; 1999) conclusion that "it is difficult to designate a specific cutoff value for each fit index because it does not work equally well with various types of fit indices, sample sizes, estimators, or distributions" (page 449). Moreover, they argued that high incremental fit indices (> 0.90) is not a sufficient basis to establish the validity of interpretations based on the theory underlying the posited model. Rather, as pointed out by Hu and Bentler (1998), "consideration of other aspects such as adequacy and interpretability of parameter estimates, model complexity, and many other issues remains critical in deciding on the validity of a model" (page 450). In their recent comment on the dangers of setting cutoff values for fit indices, Marsh et al. (2004) recommended that interpretations of model-fit "should ultimately have to be evaluated in relation to substantive and theoretical issues that are likely to be idiosyncratic to a particular study" (page 340).

15.5 Model Assessment

In model testing, the model initially specified by the researcher is not assumed to hold exactly in the population and may only be tentative. Its fit to the data is to be evaluated and assessed in relation to what is known about the substantive area, the quality of the data, and the extent to which various assumptions are satisfied. The goal is to derive a model that not only fits the data well from a statistical point of view, taking all aspects of error into account, but also has the property that every parameter of the model can be given a substantively meaningful interpretation. Jöreskog (1993) proposed examining three classes of information when evaluating a model's goodness-of-fit.

1. **Examine the parameter estimates** to see if there are any unreasonable values or other anomalies. Parameter estimates should have the right sign and size according to theory or *a priori* specifications. Examine the squared multiple correlation (SMC) for each relationship in the model. A SMC is an index of the amount of variance in

the observed indicator variable accounted for by its latent construct. As such, a SMC is a measure of the strength of linear relationship. A small SMC indicates a weak relationship and suggests that the model is not good.

2. **Examine the measures of overall fit of the model.** If any of these quantities indicate a poor fit of the data, proceed with the detailed assessment of fit in the next step.

3. **The tools for examining the fit in detail** are the *residuals, relative residuals,* and *standardized residuals;* the *modification indices;* and the *expected change.* All this information is presented in the various SEM software program outputs (e.g., LISREL, EQS, and AMOS), and can be used to identify the source of misspecification in the model, and to suggest how the model should be modified to fit the data better.

15.6 Improving Model Fit

Because the fit of most initial models is deemed unsatisfactory (Jöreskog & Sörbom, 1989), "model modification … has been an inevitable process in the application of covariance structure analysis" (Chou & Bentler, 1993, page 97). Broadly, model modification, under such circumstances, consists of freeing fixed parameters with the aim of improving the fit of the model. In most instances, parameters are freed sequentially, one at a time, until the researcher is satisfied with the fit of the revised model (Pedhazur, 1997).

15.6.1   Modification Indices

One useful aid in assessing the fit of a specified model involves **modification indices**, which are calculated for each nonestimated (fixed and constrained) parameter. Each such modification index measures how much a chi-square value is expected to decrease if a particular constrained parameter is set free (i.e., estimated) and the model is reestimated. The largest modification index tells us which parameter to set free to improve the fit maximally (Jöreskog & Sörbom, 1993). Associated with each modification index is an **expected parameter change**, which measures the magnitude and direction of change of each fixed parameter, if it is set free. This parameter differs from the modification index in that it does not indicate the change in overall model fit (χ^2), instead it depicts the change in the actual parameter value.

Although modification indices can be useful in assessing the impact of theoretically based model modifications, they should only be used to relax a parameter (with the largest modification index) if that parameter can be interpreted substantively. Model modification must have a **theoretical justification** before being considered, and even then the researcher should be

quite skeptical about the changes (MacCullum, 1986). If model respecifica-
tion is based only on the values of the modification indices, the researcher
is capitalizing on the uniqueness of these particular data, and the result will
most probably be an atheoretical but statistically significant model that has
little generalizability and limited use in testing causal relationships (Hair et
al., 1998).

15.6.2   Correlated Errors

One form of model modification is the addition of correlated errors to
improve fit. Two kinds of correlated errors can be identified: (1) those
between error terms of indicators of latent variables (i.e., measurement
errors), and (2) those between error terms of latent variables (i.e., residuals)
(Pedhazur, 1997). In cross-sectional studies, an important assumption underlying latent variable analysis is that the error terms between indicator variables are uncorrelated. If the error terms for two or more indicators correlate,
this means that these indicators measure something else or something in
addition to the construct they are supposed to measure. If this is the case,
the meaning of the construct and its dimensions may be different from what
is intended.

There are gains and drawbacks in the use of correlated errors of latent
variables to improve model fit. The virtue of adding correlated errors of
measurement is that such a move can result in dramatic improvements in
the overall fit of a model and, on occasion, can reveal unexpected and
possibly problematic sources of covariance among ratings or items of a
measure. The drawback to adding correlated errors of measurement is that
such a move is almost always post hoc and rarely eventuates in a satisfactory
explanation for the correlation. Thus, the likelihood that the correlation is
idiosyncratic to the sample and, therefore, not likely to replicate, is disturb-
ingly high (Hoyle & Smith, 1994). As Gerbing and Anderson (1984) put it,
"While the use of correlated errors improves the fit by accounting for ...
unwanted covariation, it does so at a corresponding loss of the meaning and
substantive conclusions which can be drawn from the model" (page 574).
Simply, it is a widespread misuse of structural equation modeling to include
correlated error terms (whatever the type) in the model for the sole purpose
of obtaining a better fit to the data. **Every correlation between error terms
must be justified and interpreted substantively** (Jöreskog & Sörbom, 1993).

15.7   Checklist of Requirements

- **Sample size.** As a test of model fit, the use (and validity) of the chi-
  square test is predicated on the viability of several assumptions, one

of which specifies that the "sample size is sufficiently large" (Jöreskog & Sörbom, 1993, page 122). Unfortunately, there is a lack of agreement about the meaning of "sufficiently large." Although there is no single criterion that dictates the necessary sample size, Hair et al. (1998) suggested that the absolute minimum sample size must be at least greater than the number of covariances in the input data matrix. The most appropriate minimum ratio is ten respondents per parameter, with an increase in the sample size as model complexity increases.

- **Number of indicator variables**. A major advantage of using multiple indicators in SEM is that they afford the study of relations among latent variables uncontaminated by errors of measurement in the indicators. This is predicated, among other things, on judicious choices of indicators or manifest variables (Pedhazur, 1997). The minimum number of indicators for a construct is one, and apart from the theoretical basis that should be used to select variables as indicators of a construct, there is no upper limit in terms of the number of indicators. However, as pointed out by Bentler (1980), in practice, too many indicators make it difficult, if not impossible to, fit a model to data. As a practical matter, three is the preferred minimum number of indicators, and in most cases, five to seven indicators should represent most constructs (Hair et al., 1998). In choosing the number of indicators, researchers should be guided by the axiom that it is preferable to use a relatively small number of "good" indicators than to delude oneself with a relatively large number of "poor" ones (Pedhazur, 1997).

- **Item parcels**. Instead of using individual indicators to represent latent variables, a common practice involves creating *item parcels* based on sums of responses to individual items and then using scores on these parcels in the latent variable analysis. For example, suppose that nine items were written to measure the latent variable of *resilience*. On the basis of a factor analysis of these nine measures, divide the items into three parcels, and then sum the items in each parcel to form three measured variables to operationalize the latent variable. Following the procedure described by Russell, Kahn, Spoth, and Altmaier (1998), the development of these item parcels involves the following steps:

1. Fit a one-factor model to the nine items assessing resilience.
2. Rank order items on the basis of their loadings on this factor.
3. Assign items to parcels so as to equate the average loadings of each parcel of items on the factor.

Specifically, assign items ranked 1, 5, and 9 to parcel 1; items ranked 2, 6, and 8 to parcel 2; and items ranked 3, 4, and 7 to parcel 3. With this procedure, the resulting item parcels should reflect the underlying construct of resilience to an equal degree.

The use of item parcels in latent variable analysis is supported by a number of reasons. First, responses to individual items are likely to violate the assumptions of multivariate normality that underlie the maximum likelihood estimation procedure often used in estimating structural equation models with latent variables. Second, analyses using individual items as measured indicators for the latent variables often necessitate estimating a large number of parameters (i.e., factor loadings and error terms) in fitting the model to the data. This in turn, necessitates the use of a large sample, given the recommendation that approximately ten cases per parameter be used in testing structural equation models. Finally, by using parcel rather than individual items, the results of the analysis are not likely to be distorted by idiosyncratic characteristics of individual items. One consequence of analyzing parcels rather than individual items is that the overall fit of the model to the data is improved. This effect is due to improvements in the distribution of the measured variables and to fewer parameters being estimated as a consequence of a simpler measurement model.

15.8 Assumptions

- Observations are independent of each other.
- Random sampling of respondents.
- Linearity of relationships between exogenous and endogenous variables.
- Distribution of observed variables is multivariate normal. A lack of multivariate normality is particularly troublesome because it substantially inflates the chi-square statistic and creates upward bias in critical values for determining coefficient significance (Muthuen & Kaplan, 1985; Wang, Fan, & Wilson, 1996).

15.9 Examples of Structural Equation Modeling

The following examples will be demonstrated using the software program **AMOS 5.0**. These examples assume that the researcher already has some experience with this software program, and in particular, in the use of its powerful graphical interface. For those researchers who have not used AMOS before, suffice it to say that this program is extremely easy to use and to master. Its appeal lies in its approach of using a path diagram (via a graphical user interface) to specify a model that the researcher wants to test. Drawing path diagrams to represent hypothesized models is a perfectly natural approach to structural equation modeling. For the beginner, reading **AMOS User's Guide version 5.0** (SPSS Inc., 1998) is highly recommended.

A free student version of AMOS 5.0 can be downloaded from http://amosdevelopment.com/download/. This student version contains all the graphical tools and goodness-of-fit indices found in the full version, but is limited to only eight measurement variables.

15.10   Example 1: Linear Regression with Observed Variables

Ho (1998, 1999) investigated euthanasia and the conditions under which people are most likely to support euthanasia. Items were written to represent four factors, three of which related to three conditions of suffering, and one related to the extent of support for voluntary euthanasia. These four factors, together with their representative items, are listed in the following.

- **BODY: The debilitated nature of a person's body.**

   C1: You have lost control of all your bodily functions.

   C7: The significant person in your life has lost control of all his/her bodily functions.

   C13: A person has lost control of all his/her bodily functions.

- **FAMILY: The perceived negative impact that the terminal illness of a family member has on his/her family.**

   C5: Your terminal illness will cause you to be a burden on your family.

   C11: The terminal illness of the significant person in your life will cause him/her to be a burden on his/her family.

   C17: This person's terminal illness will cause him/her to be a burden on his/her family.

- **PAIN: The experience of physical pain.**

   C3: Your terminal illness has given you continuous excruciating pain.

   C9: The terminal illness of the significant person in your life has given him/her continuous excruciating pain.

   C15: This person's terminal illness has given him/her continuous excruciating pain.

- **VOLUNTARY EUTHANASIA: The extent of support for voluntary euthanasia.**

  E1: Doctors have the right to administer medication that will painlessly end the life of a terminally ill person, if he/she requests it.

  E2: Terminally ill patients have the right to decide that life-supporting drugs or mechanisms be withheld or withdrawn, to hasten their death.

  E5: Terminally ill patients have the right to decide about their own lives and deaths.

15.10.1 Data Entry Format

The variance–covariance matrix is generated from the SPSS file: **EUTHAN.SAV**.

Variables	Column(s)	Code
E1 to E12	1–12	1 = Strongly disagree
		5 = Strongly agree
C1 to C18	13–30	1 = Strongly disagree
		5 = Strongly agree
GENDER	31	1 = Male, 2 = Female

15.10.2 Modeling in Amos Graphics

This example will demonstrate a conventional regression analysis, predicting a single observed variable (E1: extent of support for voluntary euthanasia) as a linear combination of three other observed variables (C3, C9, C15: the experience of physical pain).

- **Regression model**. Figure 15.1 presents the regression model to be tested. This model was drawn using the icons displayed in the toolbox of the **AMOS 5.0 Graphics** main window.

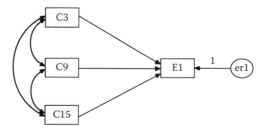

FIGURE 15.1
Conventional linear regression with observed variables.

This model evaluates how scores on C3, C9, and C15 (the experience of physical pain) are related to support for E1 (voluntary euthanasia). As it is not assumed that E1 will be perfectly predicted by a linear combination of C3, C9, and C15, the model includes an error term (er1). The variable er1 is enclosed in a circle because it is not directly observed. The single-headed arrows represent linear dependencies, and the double-headed arrows connecting the predictor variables suggest that these variables may be correlated with each other. Notice that the path coefficient from er1 to E1 is *fixed* to 1. Fixing the path coefficient to unity is necessary for the model to be identified. Without this constraint, there is not enough information to estimate the regression weight for the regression of E1 on er1, and the variance of er1 at the same time.

- **Linking the model to the data set**. Once the model has been drawn, the next step is to link the model to the data set (**EUTHAN.SAV**) to be analyzed.

1. Click ▦ from the icon toolbox (File → Data Files…) to open the **Data Files** window.

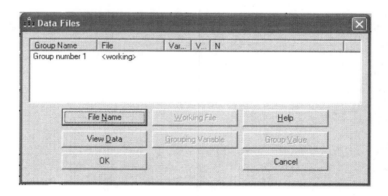

2. Click File Name to search for the data file (**EUTHAN.SAV**) in the computer's directories. Once the data file has been located, open it. By opening the data file, AMOS will automatically link the data file to the regression model, as indicated in the window. Click OK to exit the **Data Files** window.

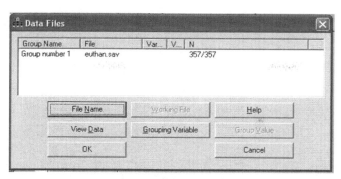

- **Results output**. To obtain specific results necessary for interpretation from the AMOS output, click the **Analysis Properties** icon 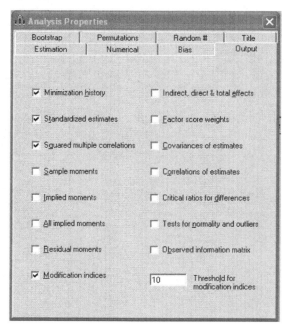 from the icon toolbox (from the menu bar, click **View/Set** → **Analysis Properties...**) to open the **Analysis Properties** window. Click the **Output** tab to open the **Output** page. Check the statistics boxes to generate the statistics required in the results output. In the present example, check the following boxes: **Minimization history, Standardized estimates, Squared multiple correlations**, and **Modification indices**. Type the number **10** in the box labeled **Threshold for modification indices** (to ensure that only those modification indices with values greater than or equal to 10 will be presented in the AMOS output). Close this window when finished.

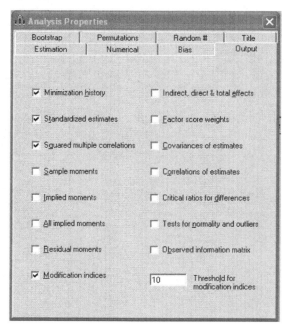

- **Conducting the analysis**. From the AMOS icon toolbox, click ▦
  (from the menu bar, click **M̲odel-Fit** → **C̲alculate Estimates**) to conduct the analysis

15.10.3   Results and Interpretation

Recall that in SEM, the goodness-of-fit of the hypothesized model is indicated
by the nonsignificant difference between the observed and predicted covariance matrices. As such, the smaller the chi-square value (nonsignificant),
the better the fit of the model. For this example, the chi-square value is 0,
which means that the fit of the model cannot be calculated (see Table 15.1).

The reason why the fit of the model cannot be calculated lies with the
number of degrees of freedom. In SEM, **degrees of freedom** are the number
of nonredundant variance and covariance moments in the input covariance
matrix *minus* the number of estimated coefficients. Each estimated coefficient
"uses up" a degree of freedom. A model can never estimate more coefficients
than the number of covariances, meaning that zero is the lower bound for
the degrees of freedom for any model. In the present example, the input
covariance matrix generated from the four observed variables (C3, C9, C15,
E1) contains four variances and six covariances, for a total of ten sample
moments. For the hypothesized model (Figure 15.1), there are three regression paths, four variances, and three covariances, for a total of ten parameters
that must be estimated. Hence, the model has zero degrees of

TABLE 15.1

Computation of Degrees of Freedom and Chi-Square Statistics for Goodness-Of-Fit

	Weights	Covariances	Variances	Means	Intercepts	Total
Parameter Summary (Group Number 1)						
Fixed	1	0	0	0	0	1
Labeled	0	0	0	0	0	0
Unlabeled	3	3	4	0	0	10
Total	4	3	4	0	0	11

Notes for Model (Default Model)
Computation of Degrees of Freedom (Default Model)

Number of distinct sample moments	10
Number of distinct parameters to be estimated	10
Degrees of freedom (10 – 10)	0

Result (default model)
Minimum was achieved.
Chi-square = .000
Degrees of freedom = 0
Probability level cannot be computed.

freedom (10 sample moments minus 10 estimated parameters), meaning there is not enough information in the observed covariance matrix to calculate the chi-square goodness-of-fit statistic. Such a model is often called **saturated** or **just identified**.

15.10.3.1 Regression Weights, Standardized Regression Weights, and Squared Multiple Correlations

Regression weights are the unstandardized coefficient estimates generated from maximum likelihood procedure (see Table 15.2).

TABLE 15.2

Regression Weights, Standardized Regression Weights, and Squared Multiple Correlations

Estimates (Group number 1 – Default model)
Scalar Estimates (Group number 1 – Default model)
Maximum Likelihood Estimates

Regression Weights: (Group Number 1 – Default Model)

			Estimate	S.E.	C.R.	P	Label
e1	<---	c3	.478	.072	6.656	***	
e1	<---	c9	−.020	.084	−.240	.810	
e1	<---	c15	.229	.082	2.791	.005	

Standardized Regression Weights: (Group Number 1 – Default Model)

			Estimate
e1	<---	c3	.457
e1	<---	c9	−.019
e1	<---	c15	−.212

Covariances: (Group Number 1 – Default Model)

			Estimate	S.E.	C.R.	P	Label
c3	<-->	c15	.884	.079	11.254	***	
c15	<-->	c9	.955	.080	11.914	***	
c3	<-->	c9	.926	.081	11.472	***	

Correlations: (Group Number 1 – Default Model)

			Estimate
c3	<-->	c15	.743
c15	<-->	c9	.814
c3	<-->	c9	.766

Variances: (Group Number 1 – Default Model)				
	Estimate	S.E.	C.R.	P	Label
c3	1.227	.092	13.342	***	
c15	1.154	.086	13.342	***	
c9	1.191	.089	13.342	***	
er1	.836	.063	13.342	***	

Squared Multiple Correlations: (Group Number 1 – Default Model)	
	Estimate
e1	.378

Associated with each estimated **unstandardized regression** coefficient (in the **Regression Weights** table) is a **standard error** (S.E.) and a **critical ratio** (C.R.) value. The **standard error** of the coefficients represents the expected variation of the estimated coefficients, and is an index of the "efficiency" of the predictor variables in predicting the endogenous variable; the smaller the S.E. the more efficient the predictor variable is. The **critical ratio** is a test of the significance of the path coefficients. Each C.R. value is obtained by dividing that parameter estimate by its respective standard error, and it is distributed approximately as z. As such, a critical ratio that is more extreme than ±1.96 indicates a significant path ($p < .05$). Based on this criterion, it can be seen that the variables C3 and C15 are highly significant predictors of E1 (C.R. = 6.65, $p < .001$; C.R. = 2.79, $p < .01$, respectively).

Standardized regression weights (β) are standardized coefficient estimates, and are independent of the units in which all variables are measured. These standardized coefficients allow the researcher to compare directly the relative relationship between each independent variable and the dependent variable. From Table 15.2, it can be seen that ratings on the two variables of C3 and C15 (both written to measure the evaluation of pain) are both significantly and positively related to E1 (extent of support for voluntary euthanasia) ($\beta = 0.46$; $\beta = 0.21$, respectively). Thus, it can be concluded that the greater the perception of pain experienced by oneself and by a nondescript person, the greater is one's support for voluntary euthanasia.

The **covariances** (unstandardized correlation coefficients) between the three predictor variables are all highly significant by the C.R. test ($p < .001$). The standardized **correlation** coefficients are all positive and ranged from 0.74 (C3<--->C15) to 0.81 (C15<--->C9).

Squared multiple correlation is an index of the proportion of the variance of the endogenous variable (E1) that is accounted for by the exogenous or predictor variables. It can be assumed that the higher the value of the squared multiple correlation, the greater the explanatory power of the regression model, and therefore the better the prediction of the dependent variable. In the present example, the predictor variables of C3, C9, and C15 accounted for 0.378 or 37.8% of the variance of E1. As such the residual or the amount of unexplained variance (er1) for this model (support for active euthanasia) is 0.622 or 62.2% (calculated as 1 – square multiple correlation).

15.11 Example 2: Regression with Unobserved (Latent) Variables

In the previous example, the regression model incorporated only observed (i.e., measured) variables. However, these observed variables (C3, C9, C15, E1) are attitudinal variables and would, therefore, be unreliable to some degree. Unreliability of predictor variables, in particular, can be problematic as it can lead to biased regression estimates. The present example demonstrates the use of unobserved (latent) variables in a similar regression analysis. The use of unobserved variables (rather than observed variables) allows the researcher to incorporate the reliabilities of the measurement variables into the regression analysis, resulting in more accurate estimates.

In the previous example, support for voluntary euthanasia was measured by one observed variable (E1), and the experience of pain (the predictor) was measured by three observed variables (C3, C9, and C15). In the present example, the latent construct of *support for voluntary euthanasia* will be measured by the three observed variables of E1, E2, and E5, and the latent construct of *pain* will be measured by the three observed variables of C3, C9, and C15 (see Section 15.10 for a description of these variables). The regression model incorporating these two latent constructs and their respective measurement indicators is presented in Figure 15.2.

This model evaluates how the experience of pain (represented by the latent construct of PAIN) predicts the extent of support for voluntary euthanasia (represented by the latent construct of VOLUNTARY EUTHANASIA). As it is not assumed that the extent of one's support for voluntary euthanasia will be perfectly predicted by one's experience of pain, this dependent variable includes a residual (z1).

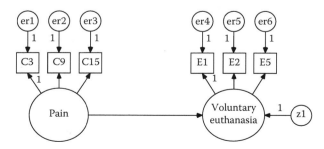

FIGURE 15.2
Regression with unobserved variables. Predicting support for voluntary euthanasia.

With nine unobserved variables in this model, it is certainly not identified. That is, there is not enough information to estimate all of the model's parameters simultaneously. For the model to be identified, it will be necessary to fix the unit of measurement of each latent variable to unity (e.g., the parameters between C3 and the latent variables of PAIN and er1 are fixed to 1). If a latent variable has more than one parameter (single-headed arrow leading away from it), then fixing anyone of them to unity will usually do. In this example, the latent variable of PAIN has three paths leading away from it, and therefore only one of these paths (between C3 and PAIN) has been constrained to unity. All paths connecting the error components (er1 to er6) to their respective indicators have also been set to unity.

The same variance–covariance matrix employed in Example 1 (generated from the SPSS file: **EUTHAN.SAV**) will be used to test the present model.

The procedure involved in (1) **Linking the model to the data set**, (2) obtaining the desired statistics in the **Results output**, and (3) **Conducting the analysis** is identical to that described in Subsection 15.10.2 (**Modeling in AMOS graphics**).

15.11.1   Results and Interpretation

The input covariance matrix generated from the model's six observed variables contains 21 sample moments. For the hypothesized model (Figure 15.2), there are five regression weights and eight variances, for a total of 13 parameters to be estimated. The model, therefore, has positive degrees of freedom $(21 - 13 = 8)$, and the chi-square goodness-of-fit statistic was computed. The result indicates that the model did not fit the data well by the chi-square test, $\chi^2(N = 357, df = 8) = 36.71$, $p < .05$ (see Table 15.3).

Although the hypothesized model did not fit the observed variance–covariance matrix well by the chi-square test, the baseline comparisons fit indices of NFI, RFI, IFI, TLI, and CFI are all above 0.9 (range: 0.949 to 0.978) (see Table 15.4). These indices compare the fit of the hypothesized model to the null or independence model. Although there are no clearly established rules as to what constitutes a good fit, a widely applied guideline for these incremental fit indices is 0.90 (Bentler, 1980; Bentler & Bonett, 1980). With the incremental fit indices ranging from 0.949 to 0.978, the possible improvement in fit for the hypothesized model (range: 0.022 to 0.051) appears so small as to be of little practical significance.

In employing the incremental fit indices to evaluate model fit, it is necessary to take into account the warning provided by Marsh et al. (2004) regarding the use of these indices as "rules of thumb". Specifically, the conventional cutoff value of > .90 may not work equally well with various types of fit indices, sample sizes, estimators, or distributions. Thus, deciding on the validity of a model must take into consideration other aspects of the model such as adequacy and interpretability of parameter estimates, model

TABLE 15.3

Computation of Degrees of Freedom and Chi-Square Statistics for Goodness-of-Fit

Notes for Group (Group number 1)
The model is recursive.
Sample size = 357

Parameter Summary (Group Number 1)						
	Weights	Covariances	Variances	Means	Intercepts	Total
Fixed	9	0	0	0	0	9
Labeled	0	0	0	0	0	0
Unlabeled	5	0	8	0	0	13
Total	14	0	8	0	0	22

Notes for Model (Default Model)
Computation of Degrees of Freedom (Default Model)

Number of distinct sample moments	21
Number of distinct parameters to be estimated	13
Degrees of freedom (21 − 13)	8

Result (default model)
Minimum was achieved.
Chi-square = 36.706
Degrees of freedom = 8
Probability level = .000

TABLE 15.4

Incremental Fit Indices

	Baseline Comparisons				
Model	NFI Delta1	RFI rho1	IFI Delta2	TLI rho2	CFI
Default model	.973	.949	.979	.960	.978
Saturated model	1.000		1.000		1.000
Independence model	.000	.000	.000	.000	.000

complexity, and most importantly, the substantive and theoretical issues underlying the posited model (Hu and Bentler, 1998; Marsh et al., 2004).

15.11.1.1 Regression Weights, Standardized Regression Weights, and Squared Multiple Correlations

In the **Regression Weights** table (see Table 15.5), the results indicate that the unstandardized regression weights are all significant by the critical ratio test ($> \pm 1.96$, $p < .05$) (except for those parameters fixed to 1). In the **Standardized Regression Weights** table, the results indicate that the experience of pain is significantly and positively related to the support for voluntary euthanasia (standardized regression weight: $\beta = 0.72$, $p < .001$). Thus, the greater the perception of pain experienced, the stronger the reported support for voluntary euthanasia. The results also indicate that the six observed mea-

TABLE 15.5

Regression Weights, Standardized Regression Weights, and Squared Multiple Correlations

			Estimate	S.E.	C.R.	P	Label
Scalar Estimates (Group Number 1 – Default Model)							
Maximum Likelihood Estimates							
Regression Weights: (Group Number 1 – Default Model)							
Voluntary_Euthanasia	<---	Pain	.722	.058	12.413	***	
C3	<---	Pain	1.000				
C9	<---	Pain	1.035	.047	21.931	***	
C15	<---	Pain	1.012	.047	21.728	***	
E1	<---	Voluntary_euthanasia	1.000				
E2	<---	Voluntary_euthanasia	.599	.044	13.517	***	
E5	<---	Voluntary_euthanasia	.787	.052	15.263	***	

Standardized Regression Weights:
(Group Number 1 – Default Model)

			Estimate
Voluntary_euthanasia	<---	Pain	.720
C3	<---	Pain	.855
C9	<---	Pain	.898
C15	<---	Pain	.892
E1	<---	Voluntary_euthanasia	.818
E2	<---	Voluntary_euthanasia	.714
E5	<---	Voluntary_euthanasia	.813

Variances: (Group Number 1 – Default Model)

	Estimate	S.E.	C.R.	P	Label
pain	.896	.091	9.820	***	
z1	.433	.059	7.321	***	
er1	.331	.033	9.978	***	
er2	.231	.029	8.081	***	
er3	.237	.028	8.417	***	
er4	.445	.053	8.438	***	
er5	.309	.028	10.882	***	
er6	.287	.033	8.612	***	

Squared Multiple Correlations:
(Group Number 1 – Default Model)

	Estimate
Voluntary_euthanasia	.519
E5	.660
E2	.510
E1	.669
C15	.795
C9	.806
C3	.730

surement variables are significantly represented by their respective latent constructs ($p < .001$).

The **squared multiple correlations** show that 0.519 or 51.9% of the variance of support for VOLUNTARY EUTHANASIA is accounted for by the variance in PAIN. The remaining 0.481 or 48.1% of the variance of support for VOLUNTARY EUTHANASIA cannot be explained by the model, and is thus attributed to the unique factor z1 (residual).

15.11.1.2 *Comparing the Latent-Construct Model (Example 2) with the Observed-Measurement Model (Example 1)*

Note that the present example employed latent constructs to test the hypothesis that the experience of pain influences the extent of support for voluntary euthanasia. This model differs from the model tested in Example 1, which employed only observed (measured) variables. Comparing the squared multiple correlation for the latent construct of "support for voluntary euthanasia" in this latent-construct model to the squared multiple correlation for the measurement variable of E1 (support for voluntary euthanasia) in the observed-variable model tested in Example 1, it can be seen that the amount of variance accounted for by this latent-construct model is greater than the amount of variance accounted for in E1 by the observed-variable model (51.9% vs. 37.8%). This demonstrates the greater explanatory power of the latent-construct analysis approach over the observed-variable analysis approach in predicting the dependent variable.

Comparing the **standard errors** associated with the parameter estimates for these two models (see Table 15.6), it can be seen that the standard error estimates obtained under the present latent-construct model are all smaller than the estimates obtained under the previous observed-variable model.

Hence, the present latent-construct model's parameter estimates are more efficient (assuming that the model is correct), and are to be preferred over the ones from the observed-variable model (Example 1).

15.12 Example 3: Multimodel Path Analysis with Latent Variables

This example demonstrates a path analysis with latent constructs to investigate the **direct** and **indirect** structural relationships between the exogenous and endogenous variables in a hypothesized path model. This example is based on the same variance–covariance matrix (generated from the SPSS file: **EUTHAN.SAV**) employed in Example 1 and Example 2. The present example, based on Ho's (1999) study, will assess the relationships between three predictor variables — (1) *the debilitated nature of one's body*, (2) *the burden placed*

TABLE 15.6

Standard Error Estimates for Latent-Construct Model and
Observed-Variable Model

Latent-Construct Model			
			S.E.
Voluntary_euthanasia	<---	Pain	.058
C3	<---	Pain	
C9	<---	Pain	.047
C15	<---	Pain	.047
E1	<---	Voluntary_euthanasia	
E2	<---	Voluntary_euthanasia	.044
E5	<---	Voluntary_euthanasia	.052

Observed-Variable Model			
			S.E.
e1	<---	c3	.072
e1	<---	c9	.084
e1	<---	c15	.082

on one's family, and (3) *the experience of physical pain* — and the dependent variable of *support for voluntary euthanasia.* The hypothesized direct and indirect relationships between the three predictor variables and the dependent variable are represented in the model presented in Figure 15.3.

The model hypothesizes that assessment of the debilitated nature of one's body (**BODY**), and the extent to which one's illness is perceived to be a burden on one's family (**FAMILY**), will be related to one's decision to support voluntary euthanasia (**VOLUNTARY EUTHANASIA**), both directly and indirectly, being mediated by one's assessment of the physical pain (**PAIN**) experienced. The description of the 12 items written to represent the four factors of **BODY** (debilitated nature of one's body — C1, C7, C13), **FAMILY** (burden on one's family — C5, C11, C17), **PAIN** (experience of physical pain — C3, C9, C15), and **VOLUNTARY EUTHANASIA** (E1, E2, E5) are presented in Section 15.10.

15.12.1   Evaluation of the Measurement Model: Confirmatory Factor Analysis (CFA)

Before evaluating the fit of the path model presented in Figure 15.3, it is necessary to define a measurement model to verify that the 12 measurement variables written to reflect the four unobserved constructs (**BODY, FAMILY, PAIN, VOLUNTARY EUTHANASIA**) do so in a reliable manner. The overall fit of a measurement model is determined by a **confirmatory factor analysis** (CFA). The fit of this model is extremely important in that *all possible latent-variable structural models are nested within it.* Obtaining a poor fit at this

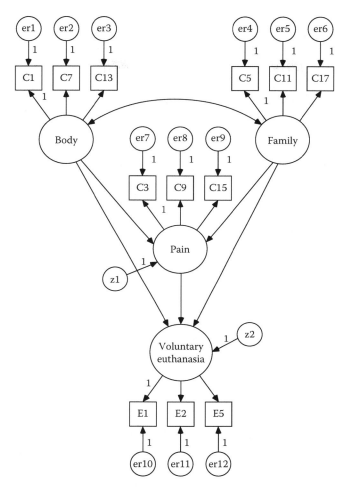

FIGURE 15.3
Path model for the prediction of voluntary euthanasia.

stage indicates a need for further refinement of the measurement model and precludes moving on to investigate latent-variable structural models (Anderson & Gerbing, 1988).

For this stage of the analysis, CFA is carried out to determine the degree of model fit, the adequacy of the factor loadings, and the standardized residuals and explained variances for the measurement variables. Figure 15.4 presents the measurement model for this example.

For this constructed measurement model, all factor loadings are freed (i.e., estimated); items are allowed to load on only one construct (i.e., no cross-loading); and latent constructs are allowed to correlate (equivalent to oblique rotation in exploratory factor analysis).

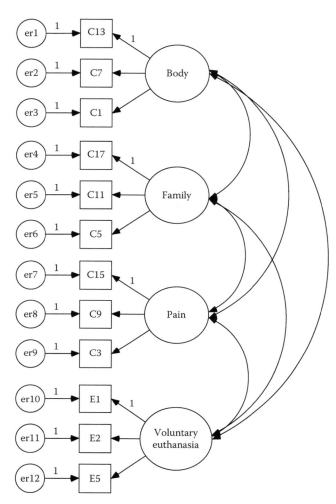

FIGURE 15.4
Measurement model.

The procedure involved in (1) **Linking the model to the data set**, (2) obtaining the desired statistics in the **Results output**, and (3) **Conducting the analysis** is identical to that described in Subsection 15.10.2.

15.12.2   Results and Interpretation

The input covariance matrix generated from the model's 12 measurement variables contains 78 sample moments. For the measurement model, there are 8 regression weights, 6 covariances, and 16 variances, for a total of 30 parameters to be estimated. The model, therefore, has 48 degrees of freedom (78 − 30), and the chi-square goodness-of-fit statistic was computed (see Table 15.7).

TABLE 15.7

Computation of Degrees of Freedom and Chi-Square Statistics for
Goodness-of-Fit

Notes for Group (Group number 1)
The model is recursive.
Sample size = 357

Parameter Summary (Group Number 1)

	Weights	Covariances	Variances	Means	Intercepts	Total
Fixed	16	0	0	0	0	16
Labeled	0	0	0	0	0	0
Unlabeled	8	6	16	0	0	30
Total	24	6	16	0	0	46

Notes for Model (Default Model)

Computation of Degrees of Freedom (Default Model)	
Number of distinct sample moments	78
Number of distinct parameters to be estimated	30
Degrees of freedom (78 − 30)	48

Result (Default model)
Minimum was achieved
Chi-square = 253.472
Degrees of freedom = 48
Probability level = .000

The chi-square goodness-of-fit test shows that the model did not fit the
data well, $\chi^2(N = 357, df = 48) = 253.472$, $p < .05$. Although the model did
not fit well by the chi-square test, the baseline comparisons fit indices of
NFI, RFI, IFI, TLI, and CFI are close to or exceed 0.9 (range: 0.89 to 0.94) (see
Table 15.8). Given the range of the computed baseline comparisons fit indices, the remaining possible improvement in fit for the hypothesized model
(range: 0.06 to 0.11) appears so small as to be of little practical significance.

TABLE 15.8

Incremental Fit Indices

	Baseline Comparisons				
Model	NFI Delta1	RFI rho1	IFI Delta2	TLI rho2	CFI
Default model	.924	.895	.937	.914	.937
Saturated model	1.000		1.000		1.000
Independence model	.000	.000	.000	.000	.000

15.12.2.1   *Regression Weights and Standardized Regression Weights*

The unstandardized regression weights are all significant by the critical ratio test ($> \pm1.96$, $p < .05$) (see Table 15.9). The standardized regression weights range from 0.706 to 0.921. These values indicate that the 12 measurement variables are significantly represented by their respective latent constructs.

Explained variances and residual variances. The explained variances for the 12 measurement variables are represented by their **squared multiple correlations** (see Table 15.10). The percentage of variance explained range from 0.498 or 49.8% (E2) to 0.848 or 84.8% (C7). The residual (unexplained) variances are calculated by subtracting each explained variance from 1 (i.e., 1 − squared multiple correlation). Thus, for the 12 measurement variables, the residual variances range from 15.2% to 50.2%.

TABLE 15.9

Regression Weights and Standardized Regression Weights

		Regression Weights: (Group Number 1 – Default Model)				
			Estimate	S.E.	C.R.	P	Label
C1	<---	BODY	1.042	.044	23.508	***	
C7	<---	BODY	1.040	.039	26.823	***	
C13	<---	BODY	1.000				
C5	<---	FAMILY	.969	.060	16.034	***	
C11	<---	FAMILY	1.025	.049	21.052	***	
C17	<---	FAMILY	1.000				
C3	<---	PAIN	.999	.045	22.183	***	
C9	<---	PAIN	1.021	.043	23.662	***	
C15	<---	PAIN	1.000				
E5	<---	Voluntary_euthanasia	.781	.050	15.499	***	
E2	<---	Voluntary_euthanasia	.588	.044	13.471	***	
E1	<---	Voluntary_euthanasia	1.000				

		Standardized Regression Weights: (Group Number 1 – Default Model)	
			Estimate
C1	<---	BODY	.862
C7	<---	BODY	.921
C13	<---	BODY	.907
C5	<---	FAMILY	.731
C11	<---	FAMILY	.898
C17	<---	FAMILY	.879
C3	<---	PAIN	.861
C9	<---	PAIN	.893
C15	<---	PAIN	.889
E5	<---	Voluntary_euthanasia	.813
E2	<---	Voluntary_euthanasia	.706
E1	<---	Voluntary_euthanasia	.824

TABLE 15.10

Explained Variances (Squared Multiple Correlations) for the 12 Measurement Variables

	Squared Multiple Correlations: (Group Number 1 – Default Model) Estimate
E1	.678
E2	.498
E5	.660
C15	.791
C9	.798
C3	.742
C17	.773
C11	.807
C5	.534
C13	.823
C7	.848
C1	.743

Modification indices. Examination of the **Modification indices** suggests that the fit of the model can be improved substantially by allowing the error terms of er1 (associated with the measurement variable **C13**) and er7 (associated with the measurement variable **C15**) to correlate (see Table 15.11).

As can be seen from the table, allowing these two error terms to correlate will reduce the chi-square value of the modified model by at least 58.981. While this is a substantial decrease (for the loss of 1 degree of freedom), the decision to implement or not to implement this modification rests with the

TABLE 15.11

Modification Indices

		Covariances: (Group Number 1 – Default Model)	M.I.	Par Change
er9	<-->	Voluntary_euthanasia	14.183	.101
er9	<-->	PAIN	10.368	.070
er4	<-->	er7	18.018	.076
er4	<-->	er8	21.856	.084
er5	<-->	er8	18.232	.075
er6	<-->	er7	10.964	.086
er6	<-->	er9	28.331	.151
er1	**<-->**	**er7**	**58.981**	**.133**
er1	<-->	er8	17.942	.074
er1	<-->	er9	10.241	.061
er1	<-->	er4	15.178	.067
er1	<-->	er5	12.440	.059
er2	<-->	er7	13.963	.064
er2	<-->	er8	16.948	.072
er2	<-->	er4	11.948	.059
er2	<-->	er5	31.138	.093
er3	<-->	er7	14.623	.081
er3	<-->	er9	23.425	.112

researcher, and in particular, on the **theoretical justification** for this modification. As mentioned earlier, without strong theoretical justification, employing the values of the modification indices to improve model fit increases the probability that the researcher is capitalizing on the uniqueness of the particular data set, and the results will most likely be atheoretical.

For the present example, the motivation to include correlated errors in a modified model is twofold. First, allowing the error terms of er1 and er7 to correlate will reduce the chi-square goodness-of-fit value substantially (i.e., improving the model fit). Although this reason does not lower the probability that the strategy may improve the fit of the model by capitalizing on chance, it does have a legitimate place in exploratory studies (Arbuckle & Wothke, 1999). Second, and more importantly, the two measurement variables (C13, C15) associated with the error terms of er1 and er7 appear to share something in common, above and beyond the latent constructs they were written to represent. Both items C13 (a person has lost control of all his/her bodily functions) and C15 (this person's terminal illness has given him/her continuous excruciating pain) appear to reflect the physical pain commonly associated with a debilitated body.

15.12.3   The Modified Model

This modification was carried out and the modified model was reestimated. Correlation between two error terms is achieved in AMOS Graphics by joining the error terms with a double-headed arrow (see Figure 15.5).

Table 15.12 presents the chi-square goodness-of-fit value, the unstandardized and standardized regression weights for the modified model.

The chi-square goodness-of-fit value for this modified model (188.008) is smaller than the chi-square value obtained for the original model (253.472). With a smaller chi-square value, the modified model, therefore, represents a better fit to the data than the original model. However, the question remains as to whether the improvement in fit represents a statistically significant improvement.

15.12.4   Comparing the Original (Default) Model against the Modified Model

In this example, a direct comparison in goodness-of-fit between the original and modified models is possible because both models are based on the same data set, and have different degrees of freedom. A test of the original model against the modified model can be obtained by subtracting the smaller chi-square value from the larger one. In this example, the comparison chi-square value is 65.464 (i.e., 253.472 − 188.008). If the original model is correctly specified, this value will have an approximate chi-square distribution with degrees of freedom equal to the difference between the degrees of freedom of the competing models. In this example, the difference in degrees of

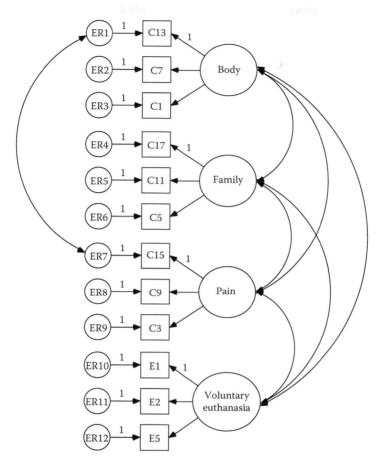

FIGURE 15.5
Modified measurement model.

freedom is 1 (i.e., 48 – 47). Therefore, with 1 degree of freedom, a chi-square value of 65.464 is significant at the 0.05 level. Based on this test, it can be concluded that the modified model (with correlated errors) represents a *significantly* better fit to the data than the original model. Please note that the regression weights, obtained under the modified model, remain statistically significant by the critical ratio test, and are highly similar to those obtained under the original model.

15.12.5   Multimodel Analysis: Evaluation of the Direct Path Model vs. the Indirect Path Model

Once the modified measurement model has been confirmed, the fit of the structural path model (with correlated error terms) (Figure 15.6) can be evaluated. The factor structure confirmed in the measurement model will

TABLE 15.12

Chi-Square Goodness-of-Fit Value, Unstandardized and Standardized Regression Weights for the Modified Measurement Model

Result (default model)
Minimum was achieved
Chi-square = 188.008
Degrees of freedom = 47
Probability level = .000
Scalar Estimates (Group number 1 – Default model)
Maximum Likelihood Estimates

Regression Weights: (Group number 1 – Default model)

			Estimate	S.E.	C.R.	P	Label
C1	<---	BODY	1.036	.043	24.302	***	
C7	<---	BODY	1.030	.037	27.659	***	
C13	<---	BODY	1.000				
C5	<---	FAMILY	.973	.061	16.014	***	
C11	<---	FAMILY	1.032	.049	21.036	***	
C17	<---	FAMILY	1.000				
C3	<---	PAIN	.986	.044	22.645	***	
C9	<---	PAIN	1.022	.041	24.897	***	
C15	<---	PAIN	1.000				
E5	<---	VOLUNTARY_EUTHANASIA	.781	.051	15.462	***	
E2	<---	VOLUNTARY_EUTHANASIA	.587	.044	13.445	***	
E1	<---	VOLUNTARY_EUTHANASIA	1.000				

Standardized Regression Weights:
(Group Number 1 – Default Model)

			Estimate
C1	<---	BODY	.867
C7	<---	BODY	.921
C13	<---	BODY	.907
C5	<---	FAMILY	.731
C11	<---	FAMILY	.901
C17	<---	FAMILY	.876
C3	<---	PAIN	.859
C9	<---	PAIN	.903
C15	<---	PAIN	.889
E5	<---	VOLUNTARY_EUTHANASIA	.812
E2	<---	VOLUNTARY_EUTHANASIA	.705
E1	<---	VOLUNTAR_EUTHANASIA	.824

be used as the foundation for the path model. That is, the four unobserved factors of **BODY, FAMILY, PAIN**, and **VOLUNTARY EUTHANASIA**, together with their respective measurement indicators, and the correlated error terms will be incorporated into the structure of the path model to be evaluated.

The posited model presented in Figure 15.6 contains two models — (1) the full **direct model**, which incorporates all identified paths linking the four factors, and (2) the **indirect model**, in which the two direct paths

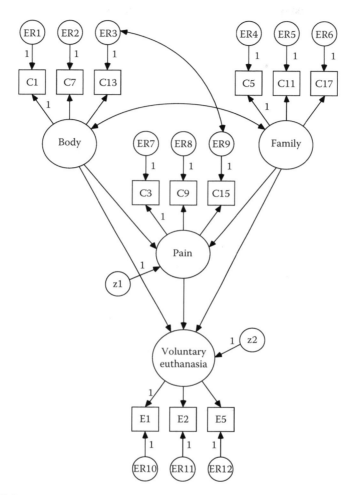

FIGURE 15.6
Path model (with correlated errors) for the prediction of voluntary euthanasia.

linking **BODY** to **VOLUNTARY EUTHANASIA** (Body → Voluntary Euthanasia) and **FAMILY** to **VOLUNTARY EUTHANASIA** (Family → Voluntary Euthanasia) will not be estimated. As both these models are nested (i.e., they are hierarchical models based on the same data set) and possess different degrees of freedom, their goodness-of-fit can be directly compared via **multimodel analysis**.

In conducting a multimodel analysis using AMOS Graphics, the procedure involves (1) defining the full direct model as presented in Figure 15.6, and (2) defining the indirect model in which the two direct paths linking **BODY** and **FAMILY** to **VOLUNTARY EUTHANASIA** (Body → Voluntary Euthanasia; Family → Voluntary Euthanasia) are constrained to zero. Constraining paths to zero is equivalent to those paths not being estimated.

15.12.6 Defining the Direct and Indirect Models

Follow these steps to define the **direct** and **indirect** models:

1. Retrieve Figure 15.6. Double-click on the **Body → Voluntary Euthanasia** path. The following **Object Properties** window will open. Label this path **VB** by clicking the **Parameters** tab, and in the **Regression weight** field, type the label **VB**.

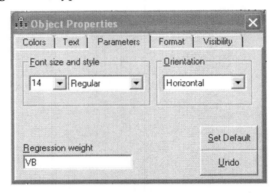

2. Double-click on the **Family → Voluntary Euthanasia** path. When the **Object Properties** window opens, click the **Parameters** tab, and in the **Regression weight** field, type the label **VF** (see Figure 15.7).

3. To define the first full **direct model**, click **Model-Fit** and then **Manage Models...** from the drop-down menu bar. When the **Manage Models** window opens, type the label **Direct Model** in the **Model Name** field. As this model imposes no constraints on any of the model's parameters, leave the **Parameter Constraints** field blank.

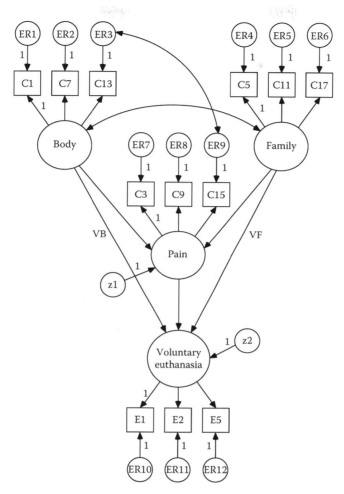

FIGURE 15.7
Direct path model vs. indirect path model.

4. To define the **indirect model**, click ▢New▢ in the **Manage Models** window. This will open a new **Manage Models** window. Type the label **Indirect Model** in the **Model Name** field. As this indirect model requires constraining the two paths of **VB** (**Body** → **Voluntary Euthanasia**) and **VF** (**Family** → **Voluntary Euthanasia**) to 0 (i.e., they will not be estimated), impose these constraints by typing **VB=0**

and **VF=0** in the **Parameter Constraints** field. Click ▢Close▢ when finished.

5. The procedure involved in (1) **Linking the two nested models to the data set**, (2) obtaining the desired statistics in the **Results output**, and (3) **Conducting the multimodel analysis** is identical to that described in Subsection 15.10.2. The multimodel analysis allows for the testing of the two nested models simultaneously, and presents a comparison of the goodness-of-fit of the two competing direct and indirect models.

15.12.7   Results and Interpretion

Summary of models. Table 15.13 presents the direct and indirect models' chi-square goodness-of-fit statistics, their baseline comparisons fit indices, and the model comparison statistics.

The direct model has 47 degrees of freedom, two less than the indirect model. This is because the direct model estimated two additional direct paths linking **BODY** and **FAMILY** to **VOLUNTARY EUTHANASIA**. The estimation of these two paths has "used up" two additional degrees of freedom. Whereas the chi-square values for both models are significant (Indirect model: $\chi^2(N = 357, df = 49) = 195.766, p < .05$; Direct model: $\chi^2(N = 357, df = 47) = 188.008, p < .05$), the baseline comparisons fit indices of NFI, RFI, IFI, TLI, and CFI for both models are above 0.90 (range: 0.921 – 0.957). These values indicate that both the hypothesized direct and indirect models fitted the observed variance–covariance matrix well relative to the null or independence model. Indeed, the only possible improvement in fit for these two models range from 0.043 to 0.079.

TABLE 15.13

Direct and Indirect Models' Chi-Square Goodness-of-Fit Indices, Baseline Comparisons Indices and Model Comparison Statistics

Model	NPAR	CMIN CMIN	DF	P	CMIN/DF
Direct model	31	188.008	47	.000	4.000
Indirect model	29	195.766	49	.000	3.995
Saturated model	78	.000	0		
Independence model	12	3332.993	66	.000	50.500

Model	NFI Delta1	RFI rho1	IFI Delta2	TLI rho2	CFI
			Baseline Comparisons		
Direct model	.944	.921	.957	.939	.957
Indirect model	.941	.921	.955	.939	.955
Saturated model	1.000		1.000		1.000
Independence model	.000	.000	.000	.000	.000

Nested Model Comparisons
Assuming Model Direct Model to Be Correct

Model	DF	CMIN	P	NFI Delta-1	IFI Delta-2	RFI rho-1	TLI rho2
Indirect model	2	7.758	.021	.002	.002	.000	.000

15.12.7.1   Goodness-of-Fit Comparison: Direct Model vs. Indirect Model

While both models fitted the data well (relative to the null model), the question arises as to which model fits better? Because both models are nested (i.e., hierarchical) and have different degrees of freedom, their goodness-of-fit can be directly compared. Looking at the **Nested Model Comparisons** statistics in Table 15.13, it can be seen that subtracting the direct model's chi-square value from the indirect model's chi-square value (195.766 − 188.008) yields a chi-square difference value of 7.758. With 2 degrees of freedom (49 − 47), this statistic is significant at the 0.05 level. Therefore, although both models fitted the data relatively well, the direct model represents a significantly better fit than the indirect model, and is to be preferred. This conclusion is further supported by the **Akaike Criterion Information** (AIC) comparison statistics. The direct model yielded a lower AIC value (250.008) than the indirect model (253.766), which indicates that the direct model is both better fitting and more parsimonious than the indirect model.

TABLE 15.14

Regression Weights, Standardized Regression Weights, and Squared Multiple Correlations for Direct Path Model

Scalar Estimates (Group Number 1 – Direct Model)
Maximum Likelihood Estimates
Regression Weights: (Group Number 1 – Direct Model)

			Estimate	S.E.	C.R.	P	Label
PAIN	<---	BODY	.380	.056	6.835	***	
PAIN	<---	FAMILY	.388	.071	5.467	***	
VOLUNTARY_ EUTHANAS	<---	PAIN	.586	.073	8.066	***	
VOLUNTARY_ EUTHANAS	<---	BODY	.135	.062	2.157	.031	VB
VOLUNTARY_ EUTHANAS	<---	FAMILY	.059	.078	.759	.448	VF
C1	<---	BODY	1.000				
C7	<---	BODY	.994	.040	24.705	***	
C13	<---	BODY	.966	.040	24.302	***	
C5	<---	FAMILY	1.000				
C11	<---	FAMILY	1.061	.065	16.282	***	
C17	<---	FAMILY	1.028	.064	16.014	***	
C3	<---	PAIN	1.000				
C9	<---	PAIN	1.037	.045	22.935	***	
C15	<---	PAIN	1.015	.045	22.645	***	
E5	<---	VOLUNTARY_ EUTHANASIA	.781	.051	15.462	***	
E2	<---	VOLUNTARY_ EUTHANASIA	.587	.044	13.445	***	
E1	<---	VOLUNTARY_ EUTHANASIA	1.000				

Standardized Regression Weights: (Group Number 1 – Direct Model)

			Estimate
PAIN	<---	BODY	.421
PAIN	<---	FAMILY	.347
VOLUNTARY_EUTHANASIA	<---	PAIN	.584
VOLUNTARY_EUTHANASIA	<---	BODY	.149
VOLUNTARY_EUTHANASIA	<---	FAMILY	.053
C1	<---	BODY	.867
C7	<---	BODY	.921
C13	<---	BODY	.907
C5	<---	FAMILY	.731
C11	<---	FAMILY	.901
C17	<---	FAMILY	.876
C3	<---	PAIN	.859
C9	<---	PAIN	.903
C15	<---	PAIN	.889
E5	<---	VOLUNTARY_EUTHANASIA	.812
E2	<---	VOLUNTARY_EUTHANASIA	.705
E1	<---	VOLUNTARY_EUTHANASIA	.824

Covariances: (Group Number 1 – Direct Model)						
			Estimate	S.E.	C.R.	P	Label
BODY	<-->	FAMILY	.570	.068	8.341	***	
ER3	<-->	ER9	.137	.019	7.105	***	

Correlations: (Group Number 1 – Direct Model)			
			Estimate
BODY	<-->	FAMILY	.636
ER3	<-->	ER9	.583

Squared Multiple Correlations: (Group Number 1 – Direct Model)	
	Estimate
PAIN	.483
VOLUNTARY_EUTHANASIA	.525
E1	.679
E2	.497
E5	.660
C15	.791
C9	.816
C3	.737
C17	.767
C11	.812
C5	.535
C13	.823
C7	.849
C1	.751

15.12.7.2 Regression Weights, Standardized Regression Weights, and Squared Multiple Correlations

Table 15.14 presents the regression weights, the standardized regression weights, and the squared multiple correlations for the hypothesized direct path model.

Of the five coefficients associated with the paths linking the model's exogenous and endogenous variables, four are significant by the critical ratio test ($>\pm1.96$, $p < .05$). The nonsignificant coefficient is associated with the path linking **FAMILY** to **VOLUNTARY EUTHANASIA**. The standardized path coefficients have been incorporated into the final direct model presented in Figure 15.8.

The results can be interpreted as follows. The perception of the debilitated nature of a person's body (**BODY**) and the negative impact that a person's terminal illness has on his/her family (**FAMILY**) is related indirectly to the support for voluntary euthanasia, being mediated by the assessment of the

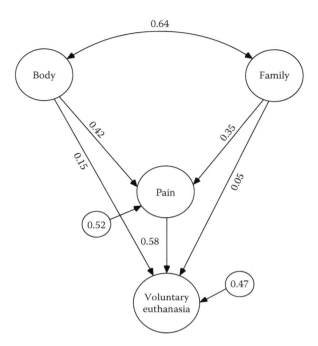

FIGURE 15.8
Direct model predicting support for voluntary euthanasia.

PAIN experienced. Thus, the more debilitated a person's body is perceived to be, and the greater the perceived negative impact on the person's family, the greater the assessment of the physical pain experienced (β = 0.42 and β = 0.35, respectively). The greater the assessment of the physical pain experienced, the greater is the support for voluntary euthanasia (β = 0.58). The perception of the debilitated nature of the body is also related directly to the support for voluntary euthanasia. Thus, the more debilitated a person's body is perceived to be, the greater is the support for voluntary euthanasia (β = 0.15). The factor of **FAMILY** was not significantly related to the support for voluntary euthanasia (p > .05).

15.12.7.3   *Explained Variances and Residual Variances*

The unidirectional arrows (without origin) pointing to the latent factors of **PAIN** and **VOLUNTARY EUTHANASIA** represent unexplained (residual) variances for these two factors. The residual variances are calculated by subtracting the factors' squared multiple correlations (explained variances) (see Table 15.14) from 1. Thus, for this hypothesized model, 52% of the variation in **PAIN** is unexplained; alternatively, 48% of the variance is accounted for by the joint influence of the **BODY** and **FAMILY** predictors.

Similarly, 47% of the variation in the support for voluntary euthanasia is unexplained; alternatively, 53% of the variance is accounted for by the joint influences of the predictors of **BODY, FAMILY,** and **PAIN**.

15.13 Example 4: Multigroup Analysis

This example demonstrates a multigroup analysis on the path model tested in Example 3. The path analysis carried out in Example 3 investigated the sequential relationships between three predictor variables — **BODY, FAMILY,** and **PAIN** — and the dependent variable of support for **VOLUNTARY EUTHANASIA**. The present example reconsiders this path model (without correlated errors) and attempts to apply it simultaneously to a sample of 136 males and a sample of 221 females. The question to be examined is whether the pattern of structural relationships hypothesized in the path model follows the same dynamics for males and females. This example is based on the same variance–covariance matrix (generated from the SPSS file: **EUTHAN.SAV**) employed in Example 1, Example 2, and Example 3.

15.13.1 Multigroup Confirmatory Factor Analysis

Prior to performing the multigroup analysis for the path model presented in Figure 15.3, it is necessary to perform a multigroup analysis for the measurement model presented in Figure 15.4. Recall that the purpose of the measurement model is to verify that the 12 measurement variables written to reflect the four latent constructs (**BODY, FAMILY, PAIN,** and **VOLUNTARY EUTHANASIA**) do so in a reliable manner. Thus, in investigating sex differences in the path model, it is necessary to first test whether the factor structure represented by the posited measurement model is the same for both males and females. *If the analysis shows no significant differences in regression weights (i.e., factor loadings) between males and females, then the same regression weights can be used for both groups.* This, in turn, will allow the regression weights themselves to be estimated more efficiently, as well as simplifying the estimation of model-fit. However, if the analysis shows significant differences in the regression weights between males and females, then these differences must be incorporated into the structural path model to be estimated.

15.13.1.1 Conducting Multigroup Modeling for Males and Females: The Measurement Model

Figure 15.4 presents the measurement model for the combined samples of males and females (N = 357). To test for sex differences in the regression weights (factor loadings) for this measurement model, it will be necessary to (1) set up separate but identical measurement models for the male and female samples, (2) link the male and female models to their respective data sets, (3) set up an invariant model (in which males and females are hypothesized to share the same regression weights) and a variant model (in which males and females are hypothesized to have different regression weights) that can be directly compared as to their model-fit, and (4) employ the **Critical Ratio** test to test for sex differences in the regression weights.

1. **Setting up separate but identical measurement models for the male and female samples:**

   To do this, retrieve the diagram presented in Figure 15.4. From the drop-down menu bar, click **Model-Fit** and then **Manage Groups...** to open the **Manage Groups** window. To create the male measurement model, change the default name (**Group number 1**) in the **Group Name** field to **Males**.

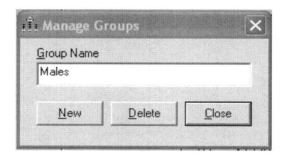

   To create the female measurement model, click **New** to open a new **Manage Groups** window. Change the name in the **Group Name** field to **Females**. Click the **Close** button. This will create two identical measurement models for the male and female samples.

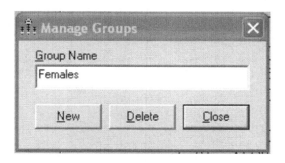

2. **Linking the male and female measurement models to their respective data sets:**

• Click 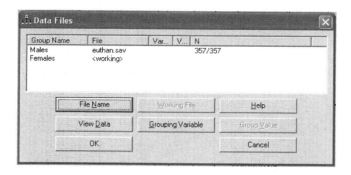 from the icon toolbox (**File → Data Files...**) to open the **Data Files** window. Click to search for the data file (**EUTHAN.SAV**) in the computer's directories. Once the data file has been located, open it. By opening the data file, AMOS will automatically link the data file to the measurement model, as indicated in the following window.

• To link the Male model to the male subsample within the **EU-THAN.SAV** data set, click to open the **Choose a Grouping Varia...** window.

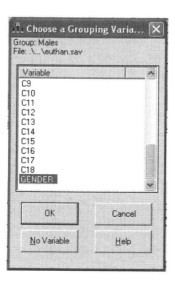

- Select **GENDER** as the grouping variable by clicking it (high-lighting), and then click ⬛ OK to return to the **Data Files** window. Notice that under the **Var...** heading, **GENDER** has been selected as the grouping variable.

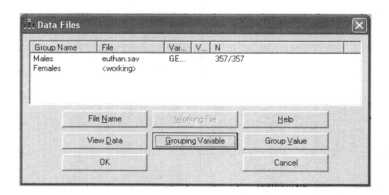

- To link the Male model to the male sub-sample (coded **1** within the **GENDER** variable), click ⬛ Group Value to open the **Choose Value for Group** window below. Select **1** (males) under the **Value** heading by clicking it (highlighting).

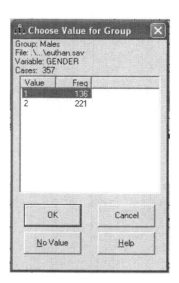

- Click 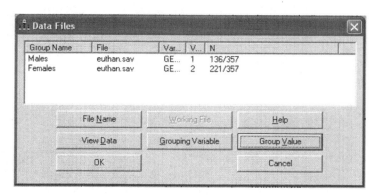 **OK** to return to the **Data Files** window. Notice that under the **V...** (**Value**) heading, the number **1** (males) has been selected. Notice also that under the **N** (sample size) heading, the number **136/357** has been listed. This indicates that a subsample of 136 males (out of a total sample of 357 subjects) has been linked to the male measurement model.

- To link the Female model to the female subsample within the **EUTHAN.SAV** data set, repeat the earlier procedure, but using the **Group Value** of **2** to select the female subsample. Successful operation of this procedure will yield the final **Data Files** window. As can be seen, a subsample of 221 females (out of a total sample of 357 subjects) has been linked to the female measurement model. Click **OK** to exit this window.

3. **Setting up group-invariant and group-variant measurement models:**

   The hypothesis to be tested is that the measurement model in Figure 15.4 holds for males as well as females. This hypothesis requires that the factor pattern (i.e., the regression weights) be the same for both groups (group-invariant model). That is, it requires that every regression weight for the male sample be *equal* to the corresponding regression weight for the female sample.

   However, it does not require that the unique variances for males and females to be group-invariant. The common factor variances and covariances may also differ in the two groups. The rationale underlying the hypothesis of group-invariant regression weights is that, while it is probably reasonable to assume that the observed and unobserved variables have different variances and covariances among males and females, the two groups may share the same regression weights.

- To set up the **group-invariant model** (in which males and females are hypothesized to share the same regression weights), it is necessary to constrain 16 paths (8 regression weights for males, 8 regression weights for females) to equality.

First, select the male measurement model by clicking the **Males** label 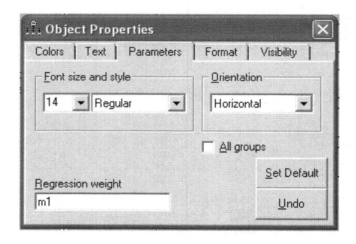 in the AMOS graphic page. Label the regression weights for this model. For example, to label the regression weight between the latent construct of **BODY** and the measurement variable of **C7** (**BODY** → C7), double-click on this parameter. This will open the **Object Properties** window. Under the **Parameters** tab, enter the label **M1** in the **Regression weight** field. Continue this procedure until all 8 regression weights (**M1** to **M8**) have been labeled (see Figure 15.9).

Second, select the female measurement model by clicking the females label in the AMOS graphic page. Repeat the above procedure for labeling the 8 regression weights for this group. For this female measurement model, label the 8 regression weights **F1** to **F8** (see Figure 15.10). That is, the labels for the regression weights for the male and female models must be different, as shown in Figure 15.9 and Figure 15.10.

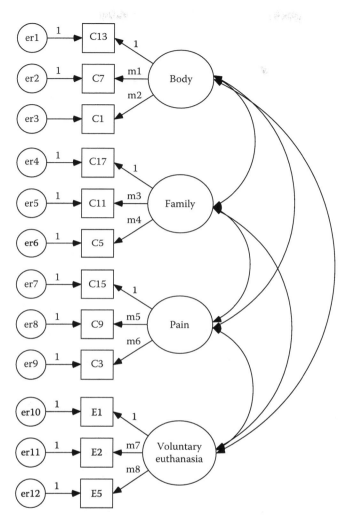

FIGURE 15.9
Measurement model — male sample.

Third, double-click on the Default model panel (**Model-Fit Manage Models...**) to open the **Manage Models** window. Set up the group-invariant model by typing in the name **group invariant** in the **Model Name** field, and then entering its constraints in the **Parameter Constraints** field (i.e., m1 = f1, m2 = f2, m3 = f3, m4 = f4, m5 = f5, m6 = f6, m7 = f7, m8 = f8). *This procedure constrains the 16 regression weights for males and females to be equivalent.*

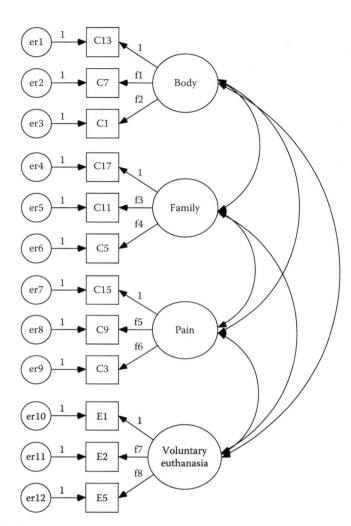

FIGURE 15.10
Measurement model — female sample.

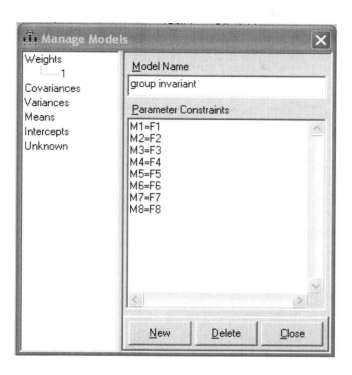

- To set up the group-variant model (in which males and females are hypothesized to have different regression weights), click

  New   in the **Manage Models** window. This will open a new **Manage Models** window. In the **Model Name** field, type the name **group variant**. As this model allows the regression weights for males and females to be estimated separately (i.e., there are no constraints), the **Parameter Constraints** field is left

  empty. Click   Close   to complete this procedure.

4. **Testing for sex differences among the regression weights:**

To test for sex differences among the regression weights, the **critical ratio** (C.R.) test can be employed to obtain the critical ratio statistics for the differences among male and female subjects' regression weights. The critical ratio for a pair of estimates provides a test of the hypothesis that the two parameters are equal (Arbuckle & Wothke, 1999).

Click the  icon (**View/Set** → **Analysis Properties...**) to open the **Analysis Properties** dialogue box. Under the **Output** tab, ensure that the **Critical ratios for differences** field is checked. This option will compare all model parameters posited for the male and female models.

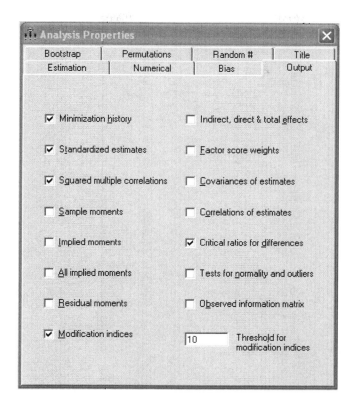

After closing the Analysis Properties dialogue box, click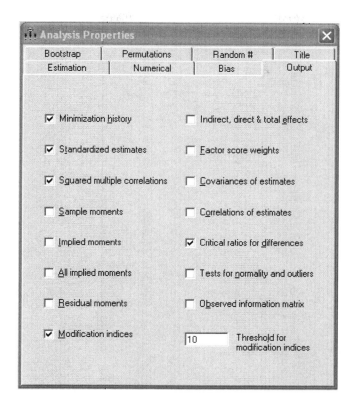
(**Model-Fit** → **Calculate Estimates**) to perform the multigroup
analysis.

15.13.1.2 Results and Interpretation

Notes for models: For this multigroup analysis, there are two data sets
(for males and females), each containing 12 measurement variables. The
two covariances matrices generated from the two data sets contain 156
sample moments.

For the **group-invariant** model, there are 52 parameters to be estimated.
This model, therefore, has 104 (156 – 52) degrees of freedom, and yielded
a significant chi-square value, $\chi^2(N = 357, df = 104) = 327.336$, $p < .05$ (see
Table 15.15).

For the **group-variant** model, there are 60 parameters to be estimated.
This model, therefore, has 96 (156 – 60) degrees of freedom, and also yielded
a significant chi-square value, $\chi^2(N = 357, df = 96) = 312.254$, $p < .05$ (see
Table 15.15).

TABLE 15.15

Computation of Degrees of Freedom and Chi-Square Goodness-of-Fit Statistics for Group-Invariant and Group-Variant Models

Notes for Model (Group Invariant) Computation of Degrees of Freedom (Group Invariant)	
Number of distinct sample moments	156
Number of distinct parameters to be estimated	52
Degrees of freedom (156 − 52)	104

Result (group invariant)
Minimum was achieved
Chi-square = 327.336
Degrees of freedom = 104
Probability level = .000

Notes for Model (Group Variant) Computation of Degrees of Freedom (Group Variant)	
Number of distinct sample moments	156
Number of distinct parameters to be estimated	60
Degrees of freedom (156 − 60)	96

Result (group variant)
Minimum was achieved
Chi-square = 312.254
Degrees of freedom = 96
Probability level = .000

Summary of models: Table 15.16 presents the chi-square goodness-of-fit statistics, baseline comparisons fit indices, and model comparison statistics for the group-invariant and group-variant measurement models.

Although the chi-square values for both models are statistically significant (i.e., both models yielded poor fit by the chi-square goodness-of-fit test), the baseline comparison fit indices of NFI, RFI, IFI, TLI, and CFI for both models are close to or are above 0.9 (range: 0.875 – 0.935). These values show the improvement in fit of both models relative to the null model. Indeed, the only possible improvement in fit for these two models ranges from 0.065 to 0.125.

Another very useful fit index is the *root mean square error of approximation* (**RMSEA**), which takes into account the error of approximation in the population. It is a measure of *discrepancy per degree of freedom* and is representative of the goodness-of-fit when the proposed model is estimated in the population. Values ranging from 0.05 to 0.08 are deemed acceptable (Browne & Cudeck, 1993; MacCallum, Browne, & Sugawara, 1996). The RMSEA values for the group-invariant and group-variant path models are 0.078 and 0.080, respectively. These values suggest that the fit of these two models is adequate.

TABLE 15.16

Chi-Square Goodness-of-Fit Statistics, Baseline Comparisons Fit
Indices and Model Comparison Statistics

	Model Fit Summary CMIN				
Model	NPAR	CMIN	DF	P	CMIN/DF
Group invariant	52	327.336	104	.000	3.147
Group variant	60	312.254	96	.000	3.253
Saturated model	156	.000	0		
Independence model	24	3442.653	132	.000	26.081

	Baseline Comparisons				
Model	NFI Delta1	RFI rho1	IFI Delta2	TLI rho2	CFI
Group invariant	.905	.879	.933	.914	.933
Group variant	.909	.875	.935	.910	.935
Saturated model	1.000		1.000		1.000
Independence model	.000	.000	.000	.000	.000

	RMSEA			
Model	RMSEA	LO 90	HI 90	PCLOSE
Group invariant	.078	.068	.087	.000
Group variant	.080	.070	.090	.000
Independence model	.266	.258	.274	.000

	AIC			
Model	AIC	BCC	BIC	CAIC
Group invariant	431.336	440.184		
Group variant	432.254	442.464		
Saturated model	312.000	338.545		
Independence model	3490.653	3494.737		

Nested Model Comparisons
Assuming Model Group Variant to Be Correct

Model	DF	CMIN	P	NFI Delta-1	IFI Delta-2	RFI rho-1	TLI rho2
Group invariant	8	15.082	.058	.004	.005	.004	.004

The fit of the two competing models can be directly compared. From the **Nested Model Comparisons** statistics, it can be seen that the chi-square difference value for the two models is 15.082 (327.336 − 312.254). With 8 degrees of freedom (104 − 96), this value is not significant at the 0.05 level ($p > .05$). Thus, the two models do not differ significantly in their goodness-of-fit.

The fit of the two models can also be compared using the **AIC** measure (Akaike, 1973, 1987). In evaluating the hypothesized model, this measure takes into account both model parsimony and model fit. Simple models that fit well receive low scores, whereas poorly fitting models get high scores. The AIC measure for the group-invariant model (431.336) is slightly lower than that for the group-variant model (432.254), indicating that the group-invariant model is both more parsimonious and better fitting than the group-variant model. On the basis of the model comparisons findings, and assuming that the group-invariant model is correct, the group-invariant model's estimates are preferable over the group-variant model's estimates.

Unstandardized regression weights and standardized regression weights: Table 15.17 presents the unstandardized regression weights and the standardized regression weights for males and females (invariant-group model).

TABLE 15.17

Unstandardized Regression Weights and Standardized Regression Weights for Males and Females

Scalar Estimates (Males – group invariant)
Maximum Likelihood Estimates

Regression Weights: (Males – Group Invariant)			Estimate	S.E.	C.R.	P	Label
C1	<---	BODY	1.039	.044	23.683	***	m2
C7	<---	BODY	1.035	.038	27.239	***	m1
C13	<---	BODY	1.000				
C5	<---	FAMILY	.964	.060	15.987	***	m4
C11	<---	FAMILY	1.031	.049	21.035	***	m3
C17	<---	FAMILY	1.000				
C3	<---	PAIN	1.015	.044	23.133	***	m6
C9	<---	PAIN	1.041	.043	24.216	***	m5
C15	<---	PAIN	1.000				
E5	<---	voluntary_euthanasia	.799	.051	15.655	***	m8
E2	<---	voluntary_euthanasia	.596	.044	13.604	***	m7
E1	<---	voluntary_euthanasia	1.000				

Standardized Regression Weights:
(Males – Group Invariant)

			Estimate
C1	<---	BODY	.855
C7	<---	BODY	.928
C13	<---	BODY	.931
C5	<---	FAMILY	.699
C11	<---	FAMILY	.899
C17	<---	FAMILY	.878
C3	<---	PAIN	.805
C9	<---	PAIN	.877
C15	<---	PAIN	.877
E5	<---	voluntary_euthanasia	.801
E2	<---	voluntary_euthanasia	.669
E1	<---	voluntary_euthanasia	.791

Scalar Estimates (Females – Group Invariant)
Maximum Likelihood Estimates

Regression Weights: (Females – Group Invariant)

			Estimate	S.E.	C.R.	P	Label
C1	<---	BODY	1.039	.044	23.683	***	m2
C7	<---	BODY	1.035	.038	27.239	***	m1
C13	<---	BODY	1.000				
C5	<---	FAMILY	.964	.060	15.987	***	m4
C11	<---	FAMILY	1.031	.049	21.035	***	m3
C17	<---	FAMILY	1.000				
C3	<---	PAIN	1.015	.044	23.133	***	m6
C9	<---	PAIN	1.041	.043	24.216	***	m5
C15	<---	PAIN	1.000				
E5	<---	voluntary_euthanasia	.799	.051	15.655	***	m8
E2	<---	voluntary_euthanasia	.596	.044	13.604	***	m7
E1	<---	voluntary_euthanasia	1.000				

Standardized Regression Weights:
(Females – Group Invariant)

			Estimate
C1	<---	BODY	.868
C7	<---	BODY	.917
C13	<---	BODY	.894
C5	<---	FAMILY	.748
C11	<---	FAMILY	.901
C17	<---	FAMILY	.876
C3	<---	PAIN	.906
C9	<---	PAIN	.913
C15	<---	PAIN	.890
E5	<---	voluntary_euthanasia	.830
E2	<---	voluntary_euthanasia	.732
E1	<---	voluntary_euthanasia	.832

The unstandardized regression weights for males and females are all significant by the critical ratio test ($> \pm 1.96$, $p < .05$). For male subjects, the standardized regression weights range from 0.669 to 0.931, and for female subjects they range from 0.732 to 0.917. These values indicate that, for both males and females, the 12 measurement variables are significantly represented by their respective unobserved constructs.

Explained variances and residual variances: For males and females, the explained variances for the 12 measurement variables are represented by their squared multiple correlations (see Table 15.18). For the male subjects, the percentage of variance explained ranged from 44.8% (E2) to 86.6% (C13); for the female subjects, the percentage of variance explained ranged from 53.6% (E2) to 84.0% (C7). The residual (unexplained) variances are calculated by subtracting each explained variance from 1. Thus, for the 12 measurement variables, the residual variances ranged from 13.4% to 55.2% for male subjects and from 16% to 46.4% for female subjects.

TABLE 15.18

Explained Variances (Squared Multiple
Correlations) for the 12 Measurement Variables

Squared Multiple Correlations: (Males — Group Invariant)	Estimate
E1	.626
E2	.448
E5	.641
C15	.769
C9	.769
C3	.648
C17	.772
C11	.808
C5	.489
C13	.866
C7	.861
C1	.731

Squared Multiple Correlations: (Females — Group Invariant)	Estimate
E1	.693
E2	.536
E5	.689
C15	.792
C9	.834
C3	.821
C17	.767
C11	.811
C5	.559
C13	.800
C7	.840
C1	.754

TABLE 15.19

Critical Ratios (C.R.) for Differences between Regression Weights
for Males and Females (Group Variant Model)

	m3	m7	m5	m4	m2	m6	m8	m1
	Pairwise Parameter Comparisons (Group Variant)							
	Critical Ratios for Differences between Parameters							
	(Group Variant)							
f3	1.600	6.179	1.996	0.817	1.246	1.855	4.470	1.244
f7	3.388	1.441	3.197	3.222	4.076	2.888	0.126	4.680
f5	1.942	6.891	**2.392**	1.034	1.587	2.193	4.994	1.636
f4	0.196	4.430	0.508	0.306	0.202	0.489	2.930	0.329
f2	1.645	6.414	2.063	0.824	1.282	1.905	4.616	1.289
f6	1.425	6.395	1.853	0.616	1.035	1.701	4.514	1.021
f8	0.790	3.761	0.492	1.140	1.273	0.436	**2.172**	1.517
f1	1.601	6.571	2.038	0.757	1.223	1.869	4.681	1.231

The critical ratio test for sex differences among the regression weights
(see Table 15.19): *Please note that the pairwise comparison C.R. test is carried out on the regression weights obtained from the variant-group model. This is because the regression weights from the invariant-group model are set to equality and therefore cannot be compared.*

From Table 15.19, it can be seen that two of the pairwise comparisons (males vs. females) for regression weights (m5 – f5, m8 – f8) are significant (C.R. > ±1.96, $p < .05$). Thus, these two sex differences in regression weights (associated with the measurement variables of **C9** and **E5**) will be incorporated into the multigroup analysis of the structural path model (see Figure 15.3).

15.13.2 Multigroup Path Analysis

Once the measurement model for both males and females has been confirmed, the fit of the structural path model posited for these two groups can be evaluated and compared. The factor structure confirmed in the measurement model will be used as the foundation for the path model. That is, the four latent constructs of **BODY, FAMILY, PAIN,** and **VOLUNTARY EUTHANASIA**, together with their respective measurement indicators, will be incorporated into the structure of the path model to be evaluated. Multigroup analysis will then be employed to apply this model simultaneously to the male and female samples. The question to be examined is whether the pattern of structural relationships hypothesized in the path model follows the same dynamics for males and females.

15.13.2.1   Conducting Multigroup Modeling for Males and Females: The Path Model

Figure 15.3 presents the original path model for the combined samples of males and females (N = 357). To test for sex differences for this path model, it will be necessary to (1) set up separate but identical path models for the male and female samples, (2) link the male and female models to their respective data sets, (3) set up an invariant path model (in which males and females are hypothesized to share the same path coefficients) and a variant path model (in which males and females are hypothesized to have different path coefficients) that can be directly compared as to their model-fit, and (4) employ the **Critical Ratio** test to test for sex differences in the path coefficients.

1. **Setting up separate but identical path models for the male and female samples:**

    To do this, retrieve the diagram presented in Figure 15.3. From the drop-down menu bar, click **Model-Fit** and then **Manage Groups...** to open the **Manage Groups** window. Change the name in the **Group Name** field to **Males**.

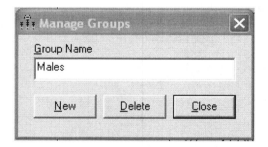

    Next, click **New** to open a new **Manage Groups** window. Change the name in the **Group Name** field to **Females**. Click the **Close** button. This will create two identical path models for the male and female samples.

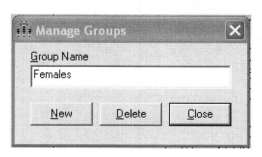

2. **Linking the male and female models to their respective data sets:**

- Click 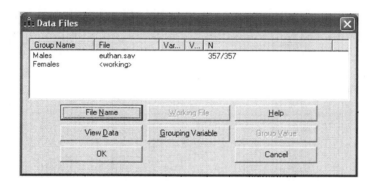 from the icon toolbox (**File → Data Files...**) to open the **Data Files** window. Click $\boxed{\text{File Name}}$ to search for the data file (**EUTHAN.SAV**) in the computer's directories. Once the data file has been located, open it. By opening the data file, AMOS will automatically link the data file to the path model, as indicated in the window.

To link the Male model to the male subsample within the **EU-THAN.SAV** data set, click $\boxed{\text{Grouping Variable}}$ to open the **Choose a Grouping Varia...** window.

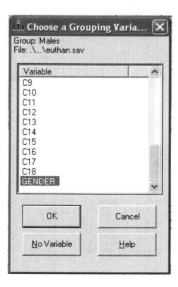

- Select **GENDER** as the grouping variable by clicking it (high-lighting), and then click 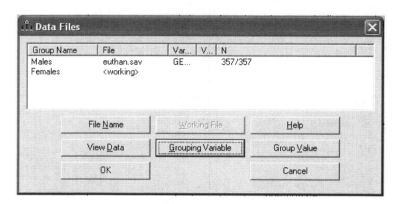 to return to the **Data Files** window. Notice that under the **Varia...** heading, **GENDER** has been selected as the grouping variable.

- To link the Male model to the male subsample (coded **1** within the **GENDER** variable), click 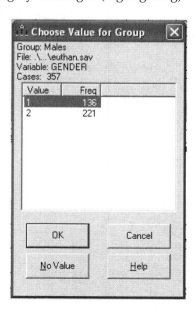 to open the **Choose Value for Group** window. Select **1** (males) under the **Value** heading by clicking it (highlighting).

- Click 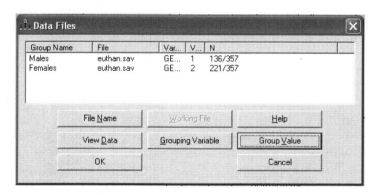 to return to the **Data Files** window. Notice that under the **V...** (**Value**) heading, the number **1** (**males**) has been selected. Notice also that under the **N** (sample size) heading, the number **136/357** has been listed. This indicates that a subsample of 136 males (out of a total sample of 357 subjects) has been linked to the male path model.

- To link the Female model to the female subsample within the **EUTHAN.SAV** data set, repeat the earlier procedure, but using the **Group Value** of **2** to select the female subsample. Successful operation of this procedure will yield the final **Data Files** window. As can be seen, a subsample of 221 females (out of a total sample of 357 subjects) has been linked to the female path model. Click [OK] to exit this window.

3. **Setting up group-invariant and group-variant path models:**

The hypothesis to be tested is that the path model in Figure 15.3 holds for both males and females. This hypothesis requires that the pattern of relationships (i.e., the path coefficients) be the same for both groups. That is, it requires that every path coefficient for the male sample be equal to the corresponding path coefficient for the female sample.

However, it does not require that the unique variances for males and females to be group-invariant. The common factor variances and covariances may also differ in the two groups. The rationale underlying the hypothesis of group-invariant path coefficients is that, although it is probably reasonable to assume that the observed and unobserved variables have different variances, covariances, and regression weights among males and females, the process by which the two groups arrived at their decision about voluntary euthanasia may be similar. If the path coefficients are the same for males and females, then the same path coefficients can be used for both groups, which simplifies the prediction of the endogenous variables from the model's exogenous variables.

- To set up the **group-invariant** path model (in which males and females are hypothesized to share the same path coefficients), it is necessary to constrain 10 paths (five path coefficients for males, five path coefficients for females) to equality.

Recall that in the prior multigroup confirmatory factor analysis (measurement) model, the critical ratio test for sex differences among the regression weights (factor loadings) yielded significant sex difference for 4 of the 16 regression weights (males and females: **C9** and **E5**) (see Table 15.9). The other 12 regression weights showed no significant sex differences. As the present path model incorporates the factor structure confirmed in the confirmatory factor analysis (measurement) model, it will therefore also be necessary to constrain these 12 regression weights to equality. That is, the two regression weights associated with the measurement variables of C9 and E5 (that showed significant sex differences) will be allowed to vary in both the group-invariant and group-variant path models (i.e., they will be estimated separately).

First, select the male path model by clicking the **Males** label

Males
Females   in the AMOS graphic page. Label the measurement variables' **regression weights** for this model by double-clicking on the first parameter to be labeled (**BODY** → C7). This will open the **Object Properties** window. Under the **Parameters** tab, enter the label **MV1** in the **Regression weight** field. Continue this procedure until all eight regression weights have been labeled (**MV1** to **MV8**) (see Figure 15.11).

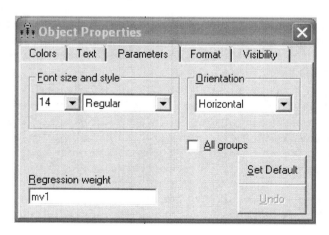

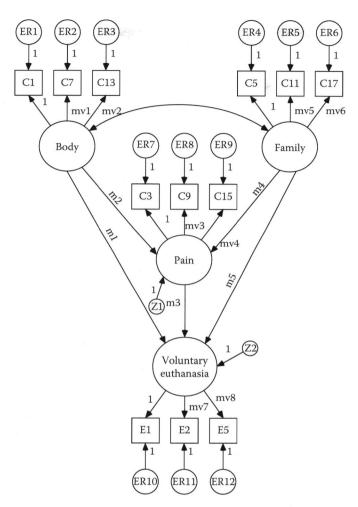

FIGURE 15.11
Male structural path model.

Second, label the **paths** for this male model. For example, to label the path between the latent construct of **BODY** and the latent construct of **VOLUNTARY EUTHANASIA** (**BODY → VOLUNTARY EUTHAN**), double-click on this parameter. This will open the **Object Properties** window. Under the **Parameters** tab, enter the label **M1** in the **Regression weight** field. Continue this procedure until all five path coefficients have been labeled (**M1** to **M5**) (see Figure 15.11).

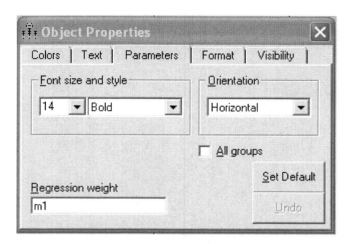

Third, select the female measurement model by clicking the females label ![Males][Females] in the AMOS graphic page. Repeat this procedure for labeling the eight regression weights and five paths for this group. For this female path model, label the eight regression weights **FV1** to **FV8** and the path coefficients **F1** to **F5** (see Figure 15.12). That is, the labels for the regression weights and the path coefficients for the male and female models must be different, as shown in Figure 15.11 and Figure 15.12.

Fourth, double-click on the ![Default model] panel (**Model-Fit Manage Models…**) to open the **Manage Models** window. Set up the group-invariant model by typing the name **group invariant** in the **Model Name** field, and then entering its constraints in the **Parameter Constraints** field. For the regression weights (factor loadings), the following 12 regression weights will be constrained to equality: **mv1=fv1, mv2=fv2, mv4=fv4, mv5=fv5, mv6=fv6, mv7=fv7** (the other four regression weights of **mv3, fv3, mv8** and **fv8** will be allowed to vary as they showed significant sex differences in the prior measurement model analysis). For the path coefficients, the following 10 paths will be constrained to equality: **m1=f1, m2=f2, m3=f3, m4=f4, m5=f5.**

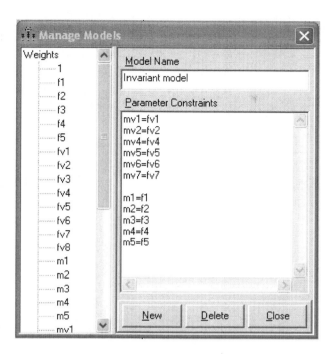

- To set up the **group-variant** model (in which males and females are hypothesized to have different path coefficients), click

  New in the **Manage Models** window. This will open a new **Manage Models** window. In the **Model Name** field, type the name **group variant**.

- As with the **group-invariant** model, this **group-variant** model requires the same 12 regression weights to be constrained to equality (**mv1=fv1, mv2=fv2, mv4=fv4, mv5=fv5, mv6=fv6, mv7=fv7**). However, this model allows the path coefficients for males and females to be estimated separately (i.e., there are no constraints). After typing the regression weights constraints in

  the **Parameter Constraints** field, click Close to complete this procedure.

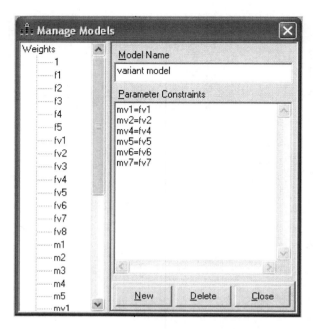

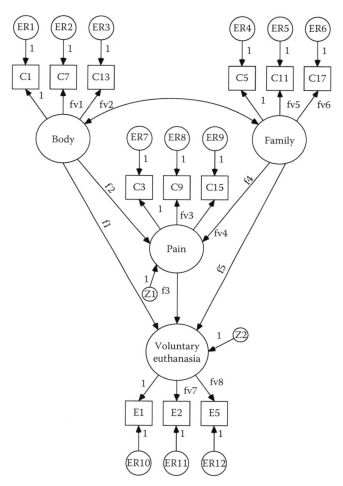

FIGURE 15.12
Female structural path model.

4. **Testing for sex differences among the path coefficients:**

To test for sex differences among the path coefficients, the **critical ratio** (C.R.) test can be employed to obtain the critical ratio statistics for the difference among male and female subjects' path coefficients.

Click the ![icon] icon (**View/Set → Analysis Properties...**) to open the **Analysis Properties** window. Under the **Output** tab, ensure that the **Critical ratios for differences** field is checked. This option will compare all model parameters posited for the male and female models.

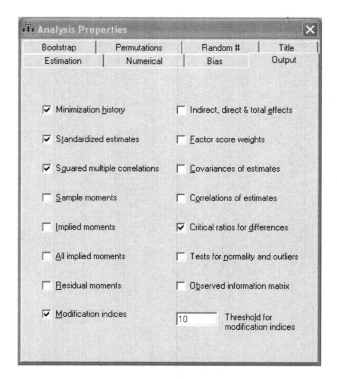

After closing the **Analysis Properties** dialog box, click 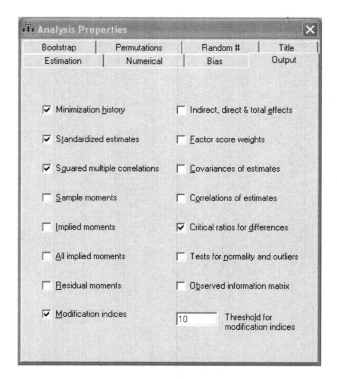 (**Model-Fit** → **Calculate Estimates**) to perform the multigroup analysis.

15.13.2.2   Results and Interpretation

Notes for models: For this multigroup analysis, there are two data sets (for males and females), each containing 12 measurement variables. The two covariance matrices generated from the two data sets contain 156 sample moments.

For the **group-invariant** model, there are 49 parameters to be estimated. This model therefore has 107 (156 − 49) degrees of freedom, and yielded a significant chi-square value, $\chi^2(N = 357, \text{df} = 107) = 327.318$, $p < .05$ (see Table 15.20).

For the **group-variant** model, there are 54 parameters to be estimated. This model, therefore, has 102 (156 − 54) degrees of freedom, and also yielded a significant chi-square value, $\chi^2(N = 357, \text{df} = 102) = 322.149$, $p < .05$ (see Table 15.20).

TABLE 15.20

Computation of Degrees of Freedom and Chi-Square
Goodness-of-Fit Statistics for Group-Invariant and Group-
Variant Path Models

Notes for Model (Group-Invariant Model)
Computation of Degrees of Freedom (Group-Invariant Model)

Number of distinct sample moments:	156
Number of distinct parameters to be estimated:	49
Degrees of freedom (156 – 49):	107

Result (group-invariant model)
Minimum was achieved
Chi-square = 327.318
Degrees of freedom = 107
Probability level = .000

Summary of models: Table 15.21 presents the chi-square goodness-of-fit
statistics, baseline comparisons fit indices, and model comparison statistics
for the group-invariant and group-variant path models.

Although the chi-square values for both path models are statistically significant (i.e., both models yielded poor fit by the chi-square goodness-of-fit
test), the baseline comparison fit indices of NFI, RFI, IFI, TLI, and CFI for
both models are close to or are above 0.90 (range: 0.879 0.934). These values
show the improvement in fit of both models relative to the null model.
Indeed, the only possible improvement in fit for these two models ranges
from 0.066 to 0.121.

The *root mean square error of approximation* (RMSEA) fit index, which takes
into account the error of approximation in the population, yielded values
for the group-invariant and group-variant path models of 0.078 and 0.076,
respectively. Values ranging from 0.05 to 0.08 are deemed acceptable (Browne
& Cudeck, 1993; MacCallum, Browne, & Sugawara, 1996). Thus, the RMSEA
values for the group-invariant and group-variant path models suggest that
the fit of these two models is adequate.

The fit of the two competing models can be directly compared. From the
Nested Model Comparisons statistics, it can be seen that the chi-square
difference value for the two models is 5.169 (327.318 – 322.149). With 5
degrees of freedom (107 – 102), this value is not significant at the 0.05 level
($p > .05$). Thus, the two models do not differ significantly in their goodness-
of-fit.

TABLE 15.21

Chi-Square Goodness-of-Fit Statistics, Baseline Comparisons Fit Indices, and Model Comparison Statistics

Model Fit Summary

Model	CMIN				
	NPAR	CMIN	DF	P	CMIN/DF
Invariant model	49	327.318	107	.000	3.059
Variant model	54	322.149	102	.000	3.158
Saturated model	156	.000	0		
Independence model	24	3442.653	132	.000	26.081

Model	Baseline Comparisons				
	NFI Delta1	RFI rho1	IFI Delta2	TLI rho2	CFI
Invariant model	.905	.883	.934	.918	.933
Variant model	.906	.879	.934	.914	.934
Saturated model	1.000		1.000		1.000
Independence model	.000	.000	.000	.000	.000

Model	RMSEA			
	RMSEA	LO 90	HI 90	PCLOSE
Invariant model	.076	.067	.086	.000
Variant model	.078	.068	.088	.000
Independence model	.266	.258	.274	.000

Model	AIC			
	AIC	BCC	BIC	CAIC
Invariant model	425.318	433.656		
Variant model	430.149	439.338		
Saturated model	312.000	338.545		
Independence model	3490.653	3494.737		

Nested Model Comparisons

Model	Assuming Model Variant Model to Be Correct						
	DF	CMIN	P	NFI Delta-1	IFI Delta-2	RFI rho-1	TLI rho2
Invariant model	5	5.169	.396	.002	.002	−.004	−.004

The fit of the two models can also be compared using the AIC measure (Akaike, 1973, 1987). In evaluating the hypothesized model, this measure takes into account both model parsimony and model fit. Simple models that fit well receive low scores, whereas poorly fitting models get high scores. The AIC measure for the group-invariant model (425.318) is lower than that for the group-variant model (430.149), indicating that the group-invariant model is both more parsimonious and better fitting than the group-variant model. On the basis of the model comparisons findings, and assuming that the group-invariant model is correct, the group-invariant model's estimates are preferable over the group-variant model's estimates.

15.13.2.2.1 Unstandardized Regression Weights, Standardized Regression Weights, and Squared Multiple Correlations

Table 15.22 presents the unstandardized regression weights, standardized regression weights, and squared multiple correlations for males and females (invariant-group model).

Of the 5 coefficients associated with the paths linking each sex-based model's exogenous and endogenous variables, 3 are significant by the critical ratio test (> ±.96, p < .05). The 2 nonsignificant coefficients for both males and females are associated with the direct paths linking **BODY → VOLUNTARY EUTHANASIA** and **FAMILY → VOLUNTARY EUTHANASIA**. These path coefficients for males and females have been incorporated into the models presented in Figure 15.13 and Figure 15.14, respectively.

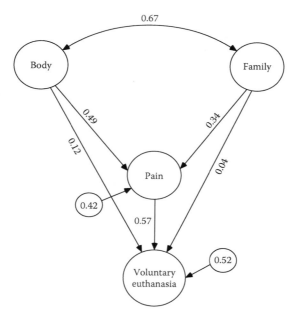

FIGURE 15.13
Male structural path model with standardized path coefficients.

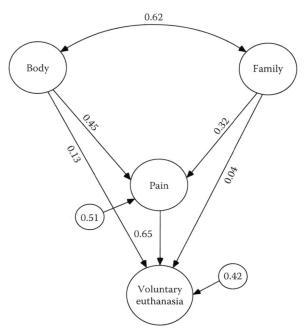

FIGURE 15.14
Female structural path model with standardized path coefficients.

The **critical ratios for differences** test for sex differences among the path coefficients (see Table 15.23). *Please note that the pairwise comparison C.R. test is carried out on the path coefficients obtained from the variant-group model. This is because the path coefficients from the invariant-group model are set to equality and, therefore, cannot be compared.*

From Table 15.23, it can be seen that none of the pairwise comparisons between path coefficients for males and females is significant ($< \pm 1.96$, $p > .05$). Thus, the hypothesized structural relationships between the three predictor variables - **BODY, FAMILY, PAIN** – and support for voluntary euthanasia, operated similarly for the male and female samples. Based on these findings, it can be concluded that for males and females, their perception of the debilitated nature of a person's body (**BODY**) and the negative impact that a person's terminal illness has on his/her family (**FAMILY**) are related indirectly to their support for voluntary euthanasia, being mediated by their assessment of the **PAIN** factor. Thus, the more debilitated a person's body is perceived to be, and the greater the perceived negative impact on the person's family, the greater the physical pain is perceived to be experienced (Male: $\beta = 0.49$ and $\beta = 0.34$, respectively; Female: $\beta = 0.45$ and $\beta = 0.33$, respectively). The greater the physical pain is perceived to be experienced, the greater is the support for voluntary euthanasia (Male: $\beta = 0.57$; Female: $\beta = 0.65$).

The **squared multiple correlations** (SMC) (see Table 15.22) present the amount of variance in the endogenous variables accounted for by the

TABLE 15.22

Unstandardized Regression Weights, Standardized Regression Weights, and Squared Multiple Correlations for Males and Females

Scalar Estimates (Male – Invariant model)
Maximum Likelihood Estimates

Regression Weights: (Male – Invariant Model)

			Estimate	S.E.	C.R.	P	Label
PAIN	<---	BODY	.426	.056	7.561	***	m2
PAIN	<---	FAMILY	.379	.072	5.297	***	m4
VOLUNTARY_ EUTHAN	<---	BODY	.114	.064	1.789	.074	m1
VOLUNTARY_EUTHAN	<---	PAIN	.612	.073	8.335	***	m3
VOLUNTARY_EUTHAN	<---	FAMILY	.049	.076	.646	.518	m5
C1	<---	BODY	1.000				
C7	<---	BODY	.996	.041	24.154	***	mv1
C13	<---	BODY	.961	.041	23.661	***	mv2
C5	<---	FAMILY	1.000				
C11	<---	FAMILY	1.069	.066	16.214	***	mv5
C17	<---	FAMILY	1.037	.065	15.965	***	mv6
C3	<---	PAIN	1.000				
C9	<---	PAIN	.942	.071	13.314	***	mv3
C15	<---	PAIN	.982	.042	23.333	***	mv4
E5	<---	VOLUNTARY_EUTHAN	.679	.073	9.319	***	mv8
E2	<---	VOLUNTARY_EUTHAN	.588	.043	13.624	***	mv7
E1	<---	VOLUNTARY_EUTHAN	1.000				

Standardized Regression Weights: (Male – Invariant Model)

			Estimate
PAIN	<---	BODY	.489
PAIN	<---	FAMILY	.340
VOLUNTARY_EUTHAN	<---	BODY	.123
VOLUNTARY_EUTHAN	<---	PAIN	.572
VOLUNTARY_EUTHAN	<---	FAMILY	.041
C1	<---	BODY	.856
C7	<---	BODY	.929
C13	<---	BODY	.930
C5	<---	FAMILY	.699
C11	<---	FAMILY	.901
C17	<---	FAMILY	.877
C3	<---	PAIN	.807
C9	<---	PAIN	.851
C15	<---	PAIN	.876
E5	<---	VOLUNTARY_EUTHAN	.759
E2	<---	VOLUNTARY_EUTHAN	.695
E1	<---	VOLUNTARY_EUTHAN	.836

Covariances: (Male – Invariant Model)

			Estimate	S.E.	C.R.	P	Label
BODY	<-->	FAMILY	.617	.108	5.711	***	par_16

<table>
<tr><td colspan="4" align="center">Correlations: (Male – Invariant model)</td></tr>
<tr><td></td><td></td><td></td><td align="right">Estimate</td></tr>
<tr><td>BODY</td><td><--></td><td>FAMILY</td><td align="right">.666</td></tr>
</table>

Squared Multiple Correlations: (Male – Invariant Model)	
	Estimate
PAIN	.576
VOLUNTARY_EUTHAN	.482
E1	.699
E2	.483
E5	.576
C15	.768
C9	.724
C3	.651
C17	.770
C11	.812
C5	.489
C13	.866
C7	.863
C1	.733

Scalar Estimates (Female – Invariant Model)
Maximum Likelihood Estimates

Regression Weights: (Female – Invariant Model)

			Estimate	S.E.	C.R.	P	Label
PAIN	<---	BODY	.426	.056	7.561	***	m2
PAIN	<---	FAMILY	.379	.072	5.297	***	m4
VOLUNTARY_EUTHAN	<---	BODY	.114	.064	1.789	.074	m1
VOLUNTARY_EUTHAN	<---	PAIN	.612	.073	8.335	***	m3
VOLUNTARY_EUTHAN	<---	FAMILY	.049	.076	.646	.518	m5
C1	<---	BODY	1.000				
C7	<---	BODY	.996	.041	24.154	***	mv1
C13	<---	BODY	.961	.041	23.661	***	mv2
C5	<---	FAMILY	1.000				
C11	<---	FAMILY	1.069	.066	16.214	***	mv5
C17	<---	FAMILY	1.037	.065	15.965	***	mv6
C3	<---	PAIN	1.000				
C9	<---	PAIN	1.056	.047	22.656	***	fv3
C15	<---	PAIN	.982	.042	23.333	***	mv4
E5	<---	VOLUNTARY_EUTHAN	.845	.061	13.932	***	fv8
E2	<---	VOLUNTARY_EUTHAN	.588	.043	13.624	***	mv7
E1	<---	VOLUNTARY_EUTHAN	1.000				

Standardized Regression Weights: (Female – Invariant Model)

			Estimate
PAIN	<---	BODY	.448
PAIN	<---	FAMILY	.324
VOLUNTARY_EUTHAN	<---	BODY	.127
VOLUNTARY_EUTHAN	<---	PAIN	.645
VOLUNTARY_EUTHAN	<---	FAMILY	.044
C1	<---	BODY	.868
C7	<---	BODY	.917
C13	<---	BODY	.893
C5	<---	FAMILY	.747
C11	<---	FAMILY	.899
C17	<---	FAMILY	.876
C3	<---	PAIN	.908
C9	<---	PAIN	.922
C15	<---	PAIN	.892
E5	<---	VOLUNTARY_EUTHAN	.844
E2	<---	VOLUNTARY_EUTHAN	.717
E1	<---	VOLUNTARY_EUTHAN	.823

Covariances: (Female – Invariant Model)

			Estimate	S.E.	C.R.	P	Label
BODY	<-->	FAMILY	.529	.079	6.710	***	par_17

Correlations: (Female – Invariant Model)

			Estimate
BODY	<-->	FAMILY	.615

Squared Multiple Correlations: (Female – Invariant Model)

	Estimate
PAIN	.485
VOLUNTARY_EUTHAN	.581
E1	.677
E2	.514
E5	.713
C15	.796
C9	.851
C3	.824
C17	.767
C11	.809
C5	.558
C13	.798
C7	.840
C1	.753

TABLE 15.23

Critical Ratios for Differences between Path Coefficients for Males and Females

Pairwise Parameter Comparisons (Variant Model)

Critical Ratios for Differences between Parameters (Variant Model)

	m1	m2	m3	m4	m5	m6
f1	**1.202**	2.922	2.230	1.329	0.478	3.377
f2	2.638	**1.152**	0.969	0.204	0.804	1.808
f3	4.283	1.125	**0.649**	2.073	2.380	0.225
f4	2.577	0.894	0.824	**0.308**	0.865	1.525
f5	0.109	4.190	3.249	2.578	**1.590**	4.548
f6	3.616	0.243	0.007	1.303	1.719	**0.628**

exogenous variables. For males, the squared multiple correlations show that (1) 57.6% of the variance of **PAIN** is accounted for by the joint influence of **BODY** and **FAMILY**, and (2) 48.2% of the variance of **VOLUNTARY EUTHANASIA** is accounted for by the joint influence of **BODY**, **FAMILY**, and **PAIN**. For females, the squared multiple correlations show that (1) 48.5% of the variance of **PAIN** is accounted for by the joint influence of **BODY** and **FAMILY**, and (2) 58.1% of the variance of **VOLUNTARY EUTHANASIA** is accounted for by the joint influence of **BODY**, **FAMILY**, and **PAIN**.

Subtracting the above SMC values from 1 provides the **standardized residuals**. These coefficients provide an estimate of the proportion of variance in each endogenous variable not predicted by its respective model. Thus, these coefficients indicate that for the male path model, 42.4% of the variance in **PAIN** is not accounted by the joint influence of **BODY** and **FAMILY**, and 51.8% of the variance in **VOLUNTARY EUTHANASIA** is not accounted for by the joint influence of **BODY**, **FAMILY**, and **PAIN**. For the female path model, 51.5% of the variance in **PAIN** is not accounted by the joint influence of **BODY** and **FAMILY**, and 41.9% of the variance in **VOL-UNTARY EUTHANASIA** is not accounted for by the joint influence of **BODY**, **FAMILY**, and **PAIN**.

16

Nonparametric Tests

16.1 Aim

Most of the tests covered in this book (e.g., t-tests and analysis of variance) are **parametric** tests in that they depend considerably on population characteristics, or parameters, for their use. The t-test, for instance, uses the sample's mean and standard deviation statistics to estimate the values of the population parameters. Parametric tests also assume that the scores being analyzed come from populations that are *normally distributed* and have *equal variances*. In practice, however, the data collected may violate one or both of these assumptions. Yet, because parametric inference tests are thought to be robust with regard to violations of underlying assumptions, researchers may use these tests even if the assumptions are not met.

When the data collected flagrantly violate these assumptions, the researcher must select an appropriate *nonparametric* test. Nonparametric inference tests have fewer requirements or assumptions about population characteristics. For example, to use these tests, it is not necessary to know the mean, standard deviation, or shape of the population scores. Because nonparametric tests make no assumptions about the form of the populations from which the test samples were drawn, they are often referred to as *distribution-free* tests.

The following nonparametric techniques will be demonstrated in this chapter:

- Chi-square (χ^2) test for single-variable experiments
- Chi-square (χ^2) test of independence between two variables
- Mann-Whitney U test for two independent samples
- Kruskal-Wallis test for several independent samples
- Wilcoxon signed ranks test for two related samples
- Friedman test for several related samples

16.2 Chi-Square (χ^2) Test for Single-Variable Experiments

The chi-square inference test is most often used with nominal data, where observations are grouped into several discrete, mutually exclusive categories, and where one counts the frequency of occurrence in each category. The single-variable chi-square test compares the observed frequencies of categories to frequencies that would be expected if the null hypothesis were true. The chi-square statistic is calculated by comparing the observed values against the expected values for each of the categories and examining the differences between them.

16.2.1 Assumptions

- Data are assumed to be a random sample.
- Independence between each observation recorded in each category. That is, each subject can only have one entry in the chi-square table.
- The expected frequency for each category should be at least 5.

16.2.2 Example 1 — Equal Expected Frequencies

Suppose a researcher is interested in determining whether there is a difference among 18-year-olds living in Rockhampton, Australia, in their preference for three different brands of cola. The researcher decides to conduct an experiment in which he randomly samples 42 18-year-olds and let them taste the three different brands. The data entered in each cell of the table in the following are the number or frequency of subjects appropriate to that cell. Thus, 10 subjects preferred Brand A; 10 subjects preferred Brand B; and 22 subjects preferred Brand C. Can the researcher conclude from these data that there is a difference in preference in the population?

Brand A	Brand B	Brand C	Total
10	10	22	42

16.2.2.1 *Data Entry Format*

The data set has been saved under the name: **EX16a.SAV**.

Variables	Column(s)	Code
SEX	1	1 = male, 2 = female
COLA	2	1 = Brand A
		2 = Brand B
		3 = Brand C

16.2.2.2 Windows Method

1. From the menu bar, click **Analyze**, then **Nonparametric Tests**, and then **Chi-Square**. The following **Chi-Square Test** window will open.

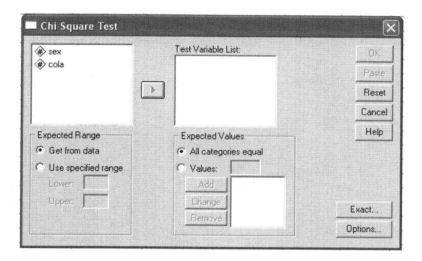

2. In the field containing the study's variables, click (highlight) the **COLA** variable, and then click 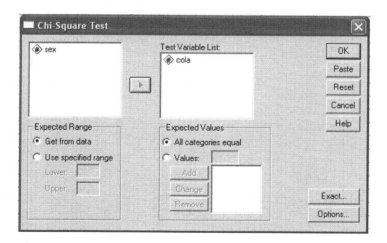 to transfer this variable to the **Test Variable List** field. Ensure that the **All categories equal** cell is checked (this is the default).

3. Click OK to run the chi-square analysis. See Table 16.1 for the results.

16.2.2.3   SPSS Syntax Method

NPAR TESTS CHISQUARE=COLA
/EXPECTED=EQUAL.

16.2.2.4   SPSS Output

TABLE 16.1

Chi-Square Output for Single-Variable (Equal Expected Frequencies)

NPar Tests
Chi-Square Test
Frequencies

Cola

	Observed N	Expected N	Residual
Brand A	10	14.0	–4.0
Brand B	10	14.0	–4.0
Brand C	22	14.0	8.0
Total	42		

Test Statistics

	Cola
Chi-Square[a]	6.857
df	2
Asymp. Sig.	.032

[a] 0 cells (.0%) have expected frequencies less than 5. The minimum expected cell frequency is 14.0.

16.2.2.5   Results and Interpretation

From the **Test Statistics** table, it can be seen that the chi-square value is significant, $\chi^2 (df = 2) = 6.86$, $p < 0.05$. There is a difference in the population regarding preference for the three brands of cola. It appears that Brand C is the favored brand.

16.2.3   Example 2 — Unequal Expected Frequencies

In the earlier example, the expected frequencies for the three categories were set to be equal (14 subjects per category). That is, the null hypothesis states that the proportion of individuals favoring Brand A in the population is equal to the proportion favoring Brand B, which is equal to the proportion favoring Brand C. This is the default null hypothesis tested. However, the researcher can change the expected frequencies in the categories to reflect certain expectations about the frequency distribution in the population. Suppose that the researcher expects that the frequency distribution of the 42 18-year-olds across the three brands of cola is not the same. More specifically, the researcher expects that 12 subjects prefer Brand A, 13 prefer Brand B, and 17 prefer Brand C. The research question then is whether there is a difference between the observations and what the researcher expects.

16.2.3.1   Windows Method

1. From the menu bar, click **Analyze**, then **Nonparametric Tests**, and then **Chi-Square**. The following **Chi-Square Test** window will open.

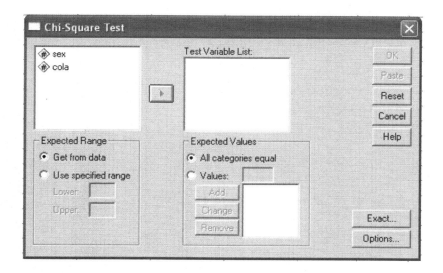

2. In the field containing the study's variables, click (highlight) the

   COLA variable, and then click 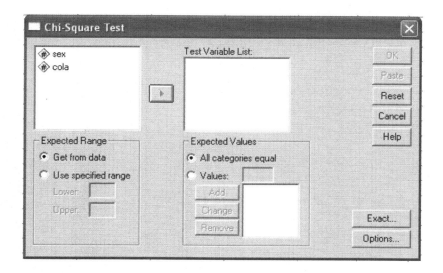 to transfer this variable to the **Test Variable List** field. Under **Expected Values**, check the **Values** cell. Type 12 (the expected frequency for Brand A) in the **Values** field. Next, type 13 (the expected frequency for Brand B) in the **Values** field. Finally, type 17 (the expected frequency for Brand C) in the **Values** field.

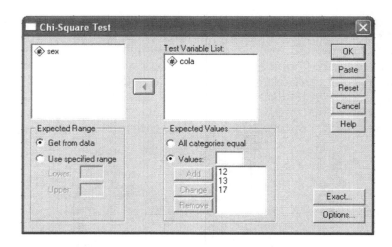

3. Click [OK] to run the chi-square analysis. See Table 16.2 for the results.

16.2.3.2   SPSS Syntax Method

NPAR TESTS CHISQUARE=COLA
/EXPECTED=12 13 17.

16.2.3.3   SPSS Output

TABLE 16.2

Chi-Square Output for Single-Variable (Unequal Expected Frequencies)

NPar Tests
Chi-Square Test
Frequencies

Cola			
	Observed N	Expected N	Residual
Brand A	10	12.0	−2.0
Brand B	10	13.0	−3.0
Brand C	22	17.0	5.0
Total	42		

Test Statistics	
	Cola
Chi-Square[a]	2.496
df	2
Asymp. Sig.	.287

[a]0 cells (.0%) have expected frequencies less than 5.
The minimum expected cell frequency is 12.0.

16.2.3.4 Results and Interpretation

From the **Test Statistics** table, it can be seen that chi-square value is not significant, $\chi^2 \,(df = 2) = 2.50$, $p > 0.05$. The observed distribution of preferences for the three brands of cola does not differ from what the researcher expected.

16.3 Chi-Square (χ^2) Test of Independence between Two Variables

The primary use of the chi-square test of independence is to determine whether two categorical variables are independent or related. To illustrate, let us suppose that the researcher in Example 1 is also interested in determining whether there is a relationship between preference for the three brands of cola and the gender of the 18-year-old subjects. The results are shown in the following 2×3 contingency table. This example will employ the data set **EX16a.SAV**.

	Brand A	Brand B	Brand C
Male	4	2	15
Female	8	11	2

16.3.1 Assumptions

- Data are assumed to be a random sample.
- Independence between each observation recorded in the contingency table, that is, each subject can only have one entry in the chi-square table.
- The expected frequency for each category should be at least 5.

16.3.2 Windows Method

1. From the menu bar, click **Analyze**, then **Descriptive Statistics**, and then **Crosstabs....** The following **Crosstabs** window will open.

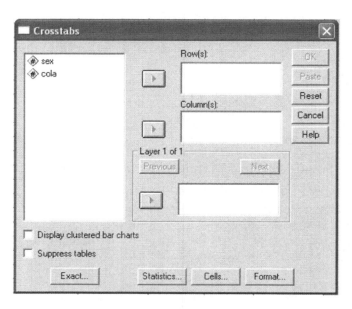

2. In the field containing the study's variables, click (highlight) the
   COLA variable, and then click [▶] to transfer this variable to the
   Row(s) field. Next, click (highlight) the **SEX** variable, and then click

   [▶] to transfer this variable to the **Column(s)** field.

3. Click 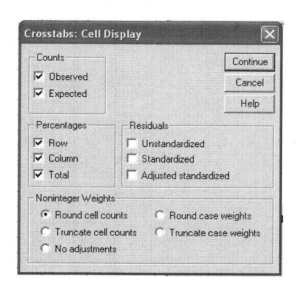 to open the **Crosstabs: Cell Display** window. Under **Counts**, check the **Observed** and **Expected** cells. Under **Per-**

centages, check the **Row, Column**, and **Total** cells. Click Continue to return to the **Crosstabs** window.

4. When the **Crosstabs** window opens, click Statistics... to open the **Crosstabs: Statistics** window. Check the **Chi-square** cell, and then

click Continue to return to the Crosstabs window.

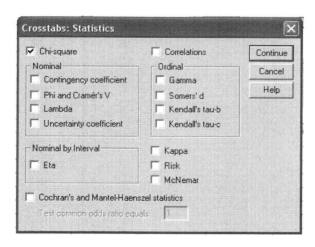

5. When the **Crosstabs** window opens, click [OK] to run the analysis. See Table 16.3 for the results.

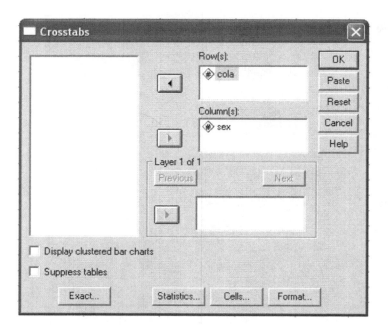

16.3.3 SPSS Syntax Method

CROSSTABS TABLES=COLA BY SEX
/CELLS=COUNT ROW COLUMN TOTAL EXPECTED
/STATISTICS≤=CHISQ.

16.3.4   SPSS Output

TABLE 16.3

Chi-Square Output for Test of Independence between Two Variables

Cola * Sex Crosstabulation

			Male	Female	Total
			Sex		
Cola	Brand A	Count	4	8	12
		Expected count	6.0	6.0	12.0
		% within cola	33.3%	66.7%	100.0%
		% within sex	19.0%	38.1%	28.6%
		% of total	9.5%	19.0%	28.6%
	Brand B	Count	2	11	13
		Expected count	6.5	6.5	13.0
		% within cola	15.4%	84.6%	100.0%
		% within sex	9.5%	52.4%	31.0%
		% of total	4.8%	26.2%	31.0%
	Brand C	Count	15	2	17
		Expected count	8.5	8.5	17.0
		% within cola	88.2%	11.8%	100.0%
		% within sex	71.4%	9.5%	40.5%
		% of total	35.7%	4.8%	40.5%
Total		Count	21	21	42
		Expected count	21.0	21.0	42.0
		% within cola	50.0%	50.0%	100.0%
		% within sex	100.0%	100.0%	100.0%
		% of total	50.0%	50.0%	100.0%

Chi-Square Tests

	Value	df	Asymp. Sig. (2-sided)
Pearson chi-square	17.505[a]	2	.000
Likelihood ratio	19.470	2	.000
Linear-by-linear Association	9.932	1	.002
N of valid cases	42		

[a]0 cells (.0%) have expected count less than 5. The minimum expected count is 6.00.

16.3.5   Results and Interpretation

The results show that the **Expected Count** frequency in each of the six cells generated by the factorial combination of SEX and COLA is greater than 5. This means that the analysis has not violated a main assumption underlying the chi-square test.

The **Pearson Chi-Square** statistic is used to determine whether there is a relationship between preference for the three brands of cola and the gender of the 18-year-old subjects. The **Pearson Chi-Square** value is statistically significant, χ^2 (df = 2) = 17.51, $p < 0.001$. This means that preference for the three brands of cola varied as a function of the subject's gender. Looking at the **Cola*Sex Cross-Tabulation** table, it can be seen that the majority of the male subjects prefer Brand C (**Count** = 15; **% Within Sex** = 71.4%) over Brand A (**Count** = 4; **% Within Sex** = 19%) and Brand B (**Count** = 2; **% Within Sex** = 9.5%). For female subjects, their preference was for Brand B (**Count** = 11; **% Within Sex** = 52.4%), followed by Brand A (**Count** = 8; **% Within Sex** = 38.1%) and Brand C (**Count** = 2; **% Within Sex** = 9.5%).

16.4 Mann–Whitney U Test for Two Independent Samples

The Mann–Whitney U test is a nonparametric test for a between-subjects design using two levels of an independent variable and scores that are measured at least at the ordinal level. It is often used in place of the t test for independent groups when there is an extreme violation of the normality assumption or when the data are scaled at a level that is not appropriate for the t test.

Suppose that the following data have been collected representing monthly incomes for teachers employed in two different private schools. The researcher wishes to determine whether there is a significant difference between these incomes.

School A		School B	
Subjects	Weekly Income ($)	Subjects	Weekly Income ($)
s1	870	s10	1310
s2	720	s11	940
s3	650	s12	770
s4	540	s13	880
s5	670	s14	1160
s6	760	s15	900
s7	730	s16	870
s8	820	s17	760
s9	1040	s18	950
		s19	1640
		s20	1270
		s21	770

16.4.1 Assumptions

- The data must be from independent, random samples.
- The data must be measured at least at the ordinal level.

- The underlying dimension of the dependent variable is continuous in nature, even though the actual measurements may be only ordinal in nature.

16.4.2 Data Entry Format

The data set has been saved under the name: **EX16b.SAV**.

Variables	Column(s)	Code
SCHOOL	1	1 = School A, 2 = School B
INCOME	2	Monthly income

16.4.3 Windows Method

1. From the menu bar, click **Analyze**, then **Nonparametric Tests**, and then **2 Independent Samples**. The following **Two-Independent-Samples Tests** window will open.

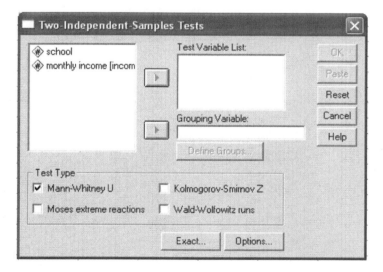

2. Because **SCHOOL** is the grouping (independent) variable, transfer it to the **Grouping Variable** field by clicking (highlighting) the variable and then clicking [▶] . As **INCOME** is the test (dependent) variable, transfer it to the **Test Variable List** field by clicking (highlighting) the variable and then clicking [▶] .

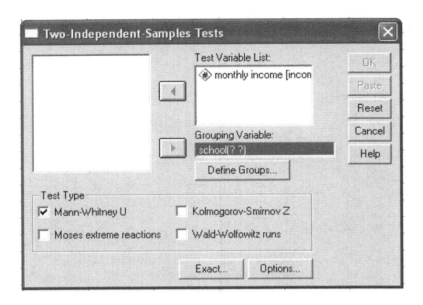

3. Click 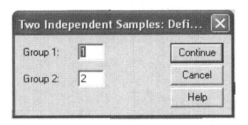 to define the ranges for the grouping variable of **SCHOOL** (coded 1 = School A, 2 = School B). When the following **Two Independent Samples: Defi...** window opens, type **1** in the Group 1 field and **2** in the Group 2 field, and then click Continue .

4. When the following **Two Independent Samples Tests** window opens, ensure that the **Mann-Whitney *U*** cell is checked. Run the analysis by clicking OK . See Table 16.4 for the results.

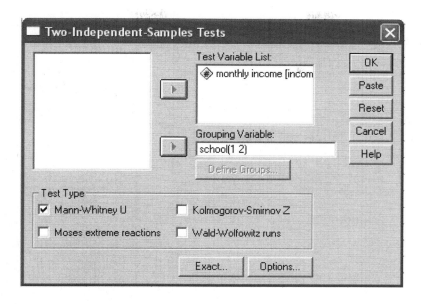

16.4.4   SPSS Syntax Method

NPAR TESTS M-W=INCOME BY SCHOOL(1,2).

16.4.5   SPSS Output

TABLE 16.4

Mann-Whitney Test Output

Mann-Whitney Test

Ranks

	School	N	Mean Rank	Sum of Ranks
Monthly income	school A	9	6.67	60.00
	school B	12	14.25	171.00
	Total	21		

Test Statistics[b]

	Monthly Income
Mann-Whitney U	15.000
Wilcoxon W	60.000
Z	−2.774
Asymp. Sig. (2-tailed)	.006
Exact Sig. [2*(1-tailed Sig.)]	.004[a]

[a]Not corrected for ties.
[b]Grouping Variable: school.

16.4.6   Results and Interpretation

The hypothesis tested by the Mann–Whitney analysis is that the *medians* of the two groups are equal. The obtained Mann–Whitney *U* statistic is 15. This value, when corrected for tied rankings and converted to a *z*-score (critical ratio test) is significant at the 0.006 level. This means that the probability of the two medians being the same is very small. Thus, it can be concluded that there is a significant difference between the median incomes of teachers in the two private schools.

16.5   Kruskal–Wallis Test for Several Independent Samples

The Kruskal–Wallis test is a nonparametric test that is used with an independent groups design comprising of more than two groups. It is a nonparametric version of the one-way ANOVA discussed in Chapter 6, and is calculated based on the sums of the ranks of the combined groups. It is used when violations of assumptions underlying parametric tests (e.g., population normality and homogeneity of variance) are extreme.

Suppose that an educational psychologist is interested in determining the effectiveness of three different methods of instruction for the basic principles of arithmetic. A total of 29 primary-grade children were randomly assigned to the three "methods-of-instruction" conditions. The scores recorded in the following are the number of correct answers obtained following completion of their instruction.

| | Instruction Method | | | | |
A		B		C	
s1	4	s10	12	s19	1
s2	5	s11	8	s20	3
s3	4	s12	10	s21	4
s4	3	s13	5	s22	6
s5	6	s14	7	s23	8
s6	10	s15	9	s24	5
s7	1	s16	14	s25	3
s8	8	s17	9	s26	2
s9	5	s18	4	s27	2

16.5.1   Assumptions

- The data must be from independent, random samples.
- The data must be measured at least at the ordinal level.
- There must be at least five scores in each sample to use the chi-square probabilities.

16.5.2 Data Entry Format

The data set has been saved under the name: **EX16c.SAV**.

Variables	Column(s)	Code
INSTRUCTION	1	1 = Method A, 2 = Method B, 3 = Method C
SCORES	2	Number of correct responses

16.5.3 Windows Method

1. From the menu bar, click **Analyze**, then **Nonparametric Tests**, and then **K Independent Samples....** The following **Tests for Several Independent Samples** window will open.

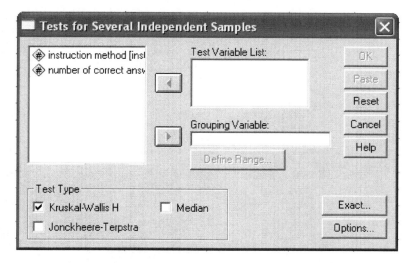

2. Because **INSTRUCTION** is the grouping (independent) variable, transfer it to the **Grouping Variable** field by clicking (highlighting) the variable and then clicking ▶ . As **SCORES** is the test (dependent) variable, transfer it to the **Test Variable List:** field by clicking (highlighting) the variable and then clicking ▶ .

3. Click **Define Ranges...** to define the ranges for the grouping variable of **INSTRUCTION** (coded 1 = Method A, 2 = Method B, and 3 = Method C). When the following **Several Independent Samples:**

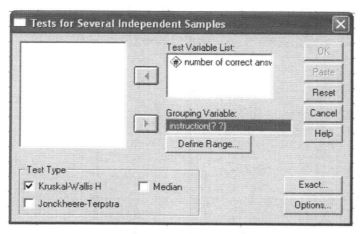

Define... window opens, type **1** in the **Minimum:** field and **3** in the

Maximum: field, and then click Continue .

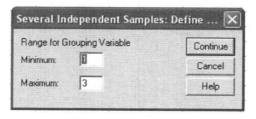

4. When the following **Tests for Several Independent Samples** window opens, ensure that the **Kruskal-Wallis H** cell is checked. Run

the analysis by clicking OK . See Table 16.5 for the results.

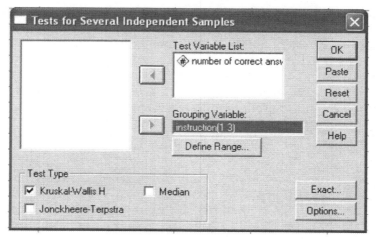

16.5.4 SPSS Syntax Method

NPAR TESTS K-W=SCORES BY INSTRUCTION(1,3).

16.5.5 SPSS Output

TABLE 16.5

Kruskal-Wallis Test Output

Kruskal-Wallis Test

Ranks

	Instruction Method	N	Mean Rank
Number of	A	9	12.72
correct answers	B	9	20.39
	C	9	8.89
	Total	27	

Test Statistics[a,b]

	Number of Correct Answers
Chi-square	9.896
df	2
Asymp. Sig.	.007

[a]Kruskal-Wallis Test.
[b]Grouping variable: instruction method.

16.5.6 Results and Interpretation

The null hypothesis tested by the Kruskal–Wallis analysis is that the three instruction methods have the same effect on the number of correct responses obtained. Therefore, the samples are random samples from the same or identical population distributions. The obtained Kruskal-Wallis statistic is interpreted as a chi-square value and is shown to be significant, χ^2 (df = 2) = 9.89, $p < 0.01$. Thus, it can be concluded that the three instruction methods are not equally effective with regard to the number of correct responses obtained.

16.6 Wilcoxon Signed Ranks Test for Two Related Samples

The Wilcoxon signed ranks test is used for within-subjects design with data that are at least ordinal in scaling. When a researcher wants to analyze two sets of data obtained from the same individuals, the appropriate test to use is the related t-test (discussed in Chapter 5). However, when there is an extreme violation of the normality assumption or when the data are not of appropriate scaling, the Wilcoxon signed ranks test can be used.

Suppose a researcher is interested in whether a drug claimed by its manufacturer to improve problem-solving skills accomplished that goal. The researcher takes a sample of seven individuals and measured their problem-solving skills before and after they have taken that drug. The number of correct responses is recorded in the following.

	Correct Responses	
	Before Drug	After Drug
s1	69	71
s2	76	75
s3	80	84
s4	74	83
s5	87	90
s6	78	75
s7	82	81

16.6.1   Assumptions

- The scale of measurement within each pair must be at least ordinal in nature.
- The differences in scores must also constitute an ordinal scale.

16.6.2   Data Entry Format

The data set has been saved under the name **EX16d.SAV**.

Variables	Column(s)	Code
BEFORE	1	Number of correct responses
AFTER	2	Number of correct responses

16.6.3   Windows Method

1. From the menu bar, click **Analyze**, then **Nonparametric Tests**, and then **2 Related Samples....** The following **Two-Related-Samples Tests** window will open.

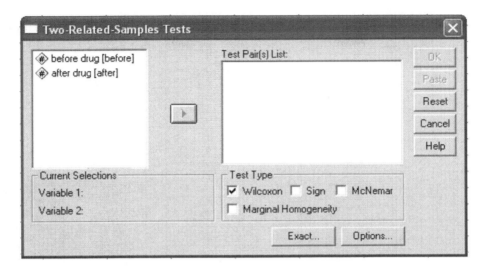

2. Transfer the two variables of **BEFORE** and **AFTER** to the **Test Pair(s)**

    List: field by clicking (highlighting) them and then clicking [▶] .
    Ensure that the **Wilcoxon** cell is checked. Run the analysis by clicking

    [OK] . See Table 16.6 for the results.

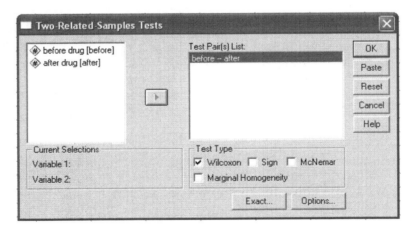

16.6.4 SPSS Syntax Method

NPAR TESTS WILCOXON=BEFORE WITH AFTER (PAIRED).

16.6.5   SPSS Output

TABLE 16.6

Wilcoxon Signed Ranks Test Output

Wilcoxon Signed Ranks Test

Ranks

		N	Mean Rank	Sum of Ranks
After drug — before drug	Negative Ranks	3[a]	2.50	7.50
	Positive Ranks	4[b]	5.13	20.50
	Ties	0[c]		
	Total			

[a]After drug < before drug.
[b]After drug > before drug.
[c]After drug = before drug.

Test Statistics[b]

	After Drug — Before Drug
z	1.103[a]
Asymp. Sig. (2-tailed)	.270

[a]Based on negative ranks.
[b]Wilcoxon Signed Ranks Test.

16.6.6   Results and Interpretation

The null hypothesis tested is that the drug does not improve problem-solving skills, i.e., it has no effect. In computing the test results, SPSS converts the Wilcoxon statistic to a z score which can be tested for significance under the normal curve. When using the 0.05 level of significance and a two-tailed test, the critical values of z are −1.96 and +1.96. Because the obtained value of z (−1.10) does not exceed these values, the researcher cannot reject the null hypothesis; i.e., it is concluded that the drug has no effect.

16.7   Friedman Test for Several Related Samples

Whereas the Wilcoxon test is used to analyze two sets of scores obtained from the same individuals, the **Friedman test** is used when there are more than two sets of scores. The Friedman test is the nonparametric alternative to a one-way repeated measures analysis of variance. Like the Mann-Whitney and Kruskal-Wallis tests, the calculation of the Friedman test is based on ranks within each case. The scores for each variable are ranked, and the mean ranks for the variables are compared.

Suppose a researcher is interested in whether problem-solving ability changes depending on the time of the day. The sample consists of 11 individuals tested in the morning, afternoon, and evening. The scores recorded in the following are the number of correct responses achieved by each subject across the three time periods.

Problem-Solving (Number of Correct Responses)

	Morning	Afternoon	Evening
s1	10	3	2
s2	6	5	15
s3	9	16	4
s4	11	12	5
s5	7	12	5
s6	12	6	11
s7	13	12	9
s8	12	3	2
s9	15	14	12
s10	13	12	5
s11	15	4	3

16.7.1 Assumptions

- The scale of measurement within the variables must be at least ordinal in nature.
- The subjects represent a random sample of subjects.

16.7.2 Data Entry Format

The data set has been saved under the name: **EX16e.SAV.**

Variables	Column(s)	Code
MORNING	1	Number of correct responses
AFTERNOON	2	Number of correct responses
EVENING	3	Number of correct responses

16.7.3 Windows Method

1. From the menu bar, click **Analyze**, then **Nonparametric Tests**, and then **K Related Samples....** The following **Tests for Several Related Samples** window will open.

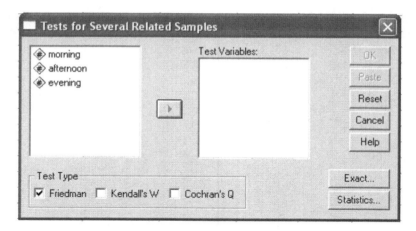

2. Transfer the three variables of **MORNING, AFTERNOON,** and **EVENING** to the **Test Variables:** field by clicking (highlighting) them and then clicking ▶ . Ensure that the **Friedman** cell is checked. Run the analysis by clicking OK . See Table 16.7 for the results.

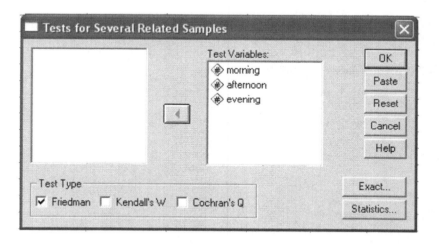

16.7.4   SPSS Syntax Method

NPAR TESTS FRIEDMAN=MORNING AFTERNOON EVENING.

16.7.5 SPSS Output

TABLE 16.7

Friedman Test Output

Friedman Test
Ranks

	Mean Rank
Morning	2.64
Afternoon	2.09
Evening	1.27

Test Statistics[a]

N	11
Chi-Square	10.364
df	2
Asymp. Sig.	.006

[a]Friedman Test.

16.7.6 Results and Interpretation

The Friedman test tests the null hypothesis that the time of the day makes no difference to the subjects' problem-solving ability (number of correct responses obtained). The results indicated that the Friedman χ^2 statistic is significant, χ^2 (df = 2) = 10.36, $p < .01$. Thus, it can be concluded that the time of the day did have a significant effect on subjects' problem-solving ability.

Appendix: Summary of SPSS Syntax Files

Frequency Analysis (Chapter 2: TRIAL.SAV)

FREQUENCIES VARIABLES=ALL *or list of variables*
/STATISTICS=ALL.

Multiple Response Analysis (Chapter 3)

- Multiple-Dichotomy Frequency Analysis (EX3.SAV)

MULT RESPONSE GROUPS=REASONS 'REASONS FOR PREFER-
RING THAT PARTY' (HONEST TO LEADER(1))
/FREQUENCIES=REASONS.

- Multiple-Response Frequency Analysis (EX3.SAV)

MULT RESPONSE GROUPS=REASONS 'REASONS FOR PREFER-
RING THAT PARTY' (REASON1 TO REASON3 (1,5))
/FREQUENCIES=REASONS.

- Multiple-Response Cross-Tabulation Frequency Analysis (EX3.SAV)

MULT RESPONSE GROUPS=REASONS 'REASONS FOR PREFER-
RING THAT PARTY' (HONEST TO LEADER(1))
/VARIABLES SEX(1,2)
/TABLES=REASONS BY SEX
/CELLS=ALL.

T-Test (Independent) (Chapter 4: EX4.SAV)

```
T-TEST GROUPS=GENDER(1,2)
/VARIABLES=WORDS.
```

T-Test (Related) (Chapter 5: EX5.SAV)

```
T-TEST PAIRS=BEFORE AFTER.
```

One-Way ANOVA (Chapter 6: EX6.SAV)

```
ONEWAY TIME BY SHOCK
/STATISTICS=DESCRIPTIVES
/RANGES=SCHEFFE(.05).
```

Factorial ANOVA (Chapter 7)

- 2 x 2 ANOVA (EX7a.SAV)
  ```
  GLM ERRORS BY STRATEGY LIST
  /EMMEANS=TABLES(STRATEGY)
  /EMMEANS=TABLES(LIST)
  /EMMEANS=TABLES(STRATEGY*LIST).
  GRAPH
  /LINE(MULTIPLE)MEAN(ERRORS) BY STRATEGY BY LIST.
  ```
- Data Transformation (Post Hoc Analysis)
  ```
  IF (STRATEGY EQ 1 AND LIST EQ 1) GROUP=1.
  IF (STRATEGY EQ 1 AND LIST EQ 2) GROUP=2.
  IF (STRATEGY EQ 2 AND LIST EQ 1) GROUP=3.
  IF (STRATEGY EQ 2 AND LIST EQ 2) GROUP=4.
  ```

VALUE LABELS GROUP 1 'A EASY' 2 'A HARD' 3 'B EASY' 4 'B HARD'.

ONEWAY ERRORS BY GROUP(1,4)

/RANGES=SCHEFFE (.05).

- 2 x 2 x 2 ANOVA (EX7b.SAV)

GLM ERRORS BY STRATEGY LIST SHOCK

/EMMEANS=TABLES(STRATEGY)

/EMMEANS=TABLES(LIST)

/EMMEANS=TABLES(SHOCK)

/EMMEANS=TABLES(STRATEGY*LIST)

/EMMEANS=TABLES(STRATEGY*SHOCK)

/EMMEANS=TABLES(LIST*SHOCK)

/EMMEANS=TABLES(STRATEGY*LIST*SHOCK.

IF (STRATEGY EQ 1 AND LIST EQ 1) GROUP=1.

IF (STRATEGY EQ 1 AND LIST EQ 2) GROUP=2.

IF (STRATEGY EQ 2 AND LIST EQ 1) GROUP=3.

IF (STRATEGY EQ 2 AND LIST EQ 2) GROUP=4.

VALUE LABELS GROUP 1 'A EASY' 2 'A HARD' 3 'B EASY' 4 'B HARD'.

GRAPH

/LINE(MULTIPLE)MEAN(ERRORS) BY STRATEGY BY LIST.

GRAPH

/LINE(MULTIPLE)MEAN(ERRORS) BY STRATEGY BY SHOCK.

GRAPH

/LINE(MULTIPLE)MEAN(ERRORS) BY LIST BY SHOCK.

GRAPH

/LINE(MULTIPLE)MEAN(ERRORS) BY SHOCK BY GROUP.

GLM Multivariate ANOVA (Chapter 8)

- 1 Sample-Test (EX8a.SAV)

COMPUTE FIFTY=(T1-6).

COMPUTE ONEHUND=(T2-12).

COMPUTE TWOHUND=(T3-25).

COMPUTE THREEHUN=(T4-40).

```
GLM FIFTY ONEHUND TWOHUND THREEHUN
/PRINT=DESCRIPTIVES.
```

- 2 Sample-Test (EX8b.SAV)

```
COMPUTE FIFTY=(T1-6).
COMPUTE ONEHUND=(T2-12).
COMPUTE TWOHUND=(T3-25).
COMPUTE THREEHUN=(T4-40).
GLM FIFTY ONEHUND TWOHUND THREEHUN BY SEX
/EMMEANS=TABLES(SEX).
```

- 2 x 2 Factorial Design (EX8c.SAV)

```
COMPUTE FIFTY=(T1-6).
COMPUTE ONEHUND=(T2-12).
COMPUTE TWOHUND=(T3-25).
COMPUTE THREEHUN=(T4-40).
GLM FIFTY ONEHUND TWOHUND THREEHUN BY SEX ETH-
    NIC
/EMMEANS=TABLES(SEX)
/EMMEANS=TABLES(ETHNIC)
/EMMEANS=TABLES(SEX*ETHNIC).
GRAPH
/LINE(MULTIPLE)MEAN(FIFTY) BY SEX BY ETHNIC.
GRAPH
/LINE(MULTIPLE)MEAN(ONEHUND) BY SEX BY ETHNIC.
GRAPH
/LINE(MULTIPLE)MEAN(TWOHUND) BY SEX BY ETHNIC.
GRAPH
/LINE(MULTIPLE)MEAN(THREEHUN) BY SEX BY ETHNIC.
IF (SEX EQ 1 AND ETHNIC EQ 1) GROUP=1.
IF (SEX EQ 1 AND ETHNIC EQ 2) GROUP=2.
IF (SEX EQ 2 AND ETHNIC EQ 1) GROUP=3.
IF (SEX EQ 2 AND ETHNIC EQ 2) GROUP=4.
VALUE LABELS GROUP 1 'MALE-WHITE' 2 'MALE-NONWHITE'
3 'FEMALE-WHITE' 4 'FEMALE-NONWHITE'.
ONEWAY ONEHUND BY GROUP (1,4)
/RANGES=SCHEFFE(.05).
```

GLM Repeated Measures Analysis (Chapter 9)

- One-Way Repeated Measures (EX9a.SAV)
  GLM TEMP1 TO TEMP4
  /WSFACTOR=TEMP 4 REPEATED
  /MEASURE=ERRORS
  /EMMEANS=TABLES(TEMP) COMPARE ADJ(BONFERRONI).
- Doubly Multivariate (EX9b.SAV)
  GLM LS_E TO HS_D
  /WSFACTOR=SHOCK 2 REPEATED ITEMS 3 REPEATED
  /MEASURE=CORRECT
  /PLOT=PROFILE(ITEMS*SHOCK)
  /EMMEANS=TABLES(SHOCK) COMPARE ADJ(BONFERRONI)
  /EMMEANS=TABLES(ITEMS) COMPARE ADJ(BONFERRONI)
  /EMMEANS=TABLES(SHOCK*ITEMS).
- 2 Factor Mixed Design (EX9c.SAV)
  GLM TRIAL1 TO TRIAL3 BY GROUP
  /WSFACTOR=TRIAL 3 REPEATED
  /MEASURE=CORRECT
  /PLOT=PROFILE(TRIAL*GROUP)
  /POSTHOC=GROUP(SCHEFFE)
  /EMMEANS=TABLES(GROUP)
  /EMMEANS=TABLES(TRIAL) COMPARE ADJ(BONFERRONI)
  /EMMEANS=TABLES(GROUP*TRIAL).
- 3 Factor Mixed Design (EX9d.SAV)
  GLM TRIAL1 TO TRIAL4 BY DRUG SEX
  /WSFACTOR=TRIAL 4 REPEATED
  /MEASURE=ERRORS
  /EMMEANS=TABLES(DRUG)
  /EMMEANS=TABLES(SEX)
  /EMMEANS=TABLES(DRUG*SEX)
  /EMMEANS=TABLES(TRIAL) COMPARE ADJ(BONFERRONI)
  /EMMEANS=TABLES(DRUG*TRIAL)
  /EMMEANS=TABLES(SEX*TRIAL)
  /EMMEANS=TABLES(DRUG*SEX*TRIAL)
  /PLOT=PROFILE(DRUG*SEX)

```
/PLOT=PROFILE(TRIAL*DRUG)
/PLOT=PROFILE(TRIAL*SEX).
IF (DRUG EQ 1 AND SEX EQ 1) GROUP=1.
IF (DRUG EQ 2 AND SEX EQ 1) GROUP=2.
IF (DRUG EQ 1 AND SEX EQ 2) GROUP=3.
IF (DRUG EQ 2 AND SEX EQ 2) GROUP=4.
VALUE LABELS GROUP 1 'DRUG PRESENT-MALE'
    2 'DRUG ABSENT-MALE'
    3 'DRUG PRESENT-FEMALE'
    4 'DRUG ABSENT-FEMALE'.
GLM TRIAL1 TO TRIAL4 BY GROUP
/WSFACTOR=TRIAL 4 REPEATED
/MEASURE=ERRORS
/PLOT=PROFILE(TRIAL*GROUP)
/EMMEANS=TABLES(GROUP*TRIAL).
```

Correlation Analysis (Chapter 10)

- Scatter Plot (CORR.SAV)
  ```
  GRAPH
  /SCATTERPLOT(BIVAR)=READ WITH GPA
  /MISSING=LISTWISE.
  ```
- Pearson Product Moment Correlation (CORR.SAV)
  ```
  CORRELATIONS READ WITH GPA
  /MISSING=PAIRWISE.
  ```
- Spearman Rank Order Correlation Coefficient (CORR.SAV)
  ```
  NONPAR CORR READ_RANK WITH GPA_RANK.
  ```

Linear Regression (Chapter 11: CORR.SAV)

```
REGRESSION VARIABLES=(COLLECT)
/STATISTICS=DEFAULTS CI
/DEPENDENT=GPA
/METHOD=ENTER READ.
```

Factor Analysis (Chapter 12: DOMES.SAV)

FACTOR VARIABLES=PROVO TO STABLE
/FORMAT=SORT BLANK(.33)
/PRINT=INITIAL EXTRACTION ROTATION KMO
/PLOT=EIGEN
/EXTRACTION=PC
/ROTATION=VARIMAX.

- Specifying Number of Factors (3) (DOMES.SAV)
  FACTOR VARIABLES=PROVO TO STABLE
  /FORMAT=SORT BLANK(.33)
  /CRITERIA=FACTOR(3)
  /EXTRACTION=PC
  /ROTATION=VARIMAX.

Reliability (Chapter 13: DOMES.SAV)

RELIABILITY VARIABLES=PROVO TO STABLE
/SCALE(SELF)=PROTECT SAVE DEFEND
/SUMMARY=TOTAL.

Multiple Regression (Chapter 14)

- Forward Selection (DOMES.SAV)
  REGRESSION VARIABLES=(COLLECT)
  /STATISTICS=DEFAULTS CHA TOL CI COLLIN
  /DEPENDENT=RESPON
  /FORWARD PROVOKE SELFDEF INSANITY.
- Hierarchical Regression (DOMES.SAV)
  REGRESSION VARIABLES=(COLLECT)
  /STATISTICS=DEFAULTS CHA TOL CI COLLIN
  /DEPENDENT=RESPON

/METHOD=ENTER SEX AGE EDUC INCOME/ENTER PROVOKE
SELFDEF INSANITY.
- Path Analysis (DOMES.SAV)
  REGRESSION VARIABLES=(COLLECT)
  /STATISTICS=DEFAULTS CHA TOL CI COLLIN
  /DEPENDENT=RESPON
  /FORWARD PROVOKE SELFDEF INSANITY.
  REGRESSION VARIABLES=(COLLECT)
  /STATISTICS=DEFAULTS CHA TOL CI COLLIN
  /DEPENDENT=SELFDEF
  /FORWARD PROVOKE INSANITY.
  REGRESSION VARIABLES=(COLLECT)
  /STATISTICS=DEFAULTS CHA TOL CI COLLIN
  /DEPENDENT=INSANITY
  /FORWARD PROVOKE.

Chi-Square (χ^2) Test for Single Variable Experiments — Equal Expected Frequencies (Chapter 16: EX16a.SAV)

NPAR TESTS CHISQUARE=COLA
/EXPECTED=EQUAL.

Chi-Square (χ^2) Test for Single Variable Experiments — Unequal Expected Frequencies (EX16a.SAV)

NPAR TESTS CHISQUARE=COLA
/EXPECTED=12 13 17.

Chi-Square (χ^2) Test of Independence between Two Variables (EX16a.SAV)

```
CROSSTABS TABLES=COLA BY SEX
/CELLS=COUNT ROW COLUMN TOTAL EXPECTED
/STATISTICS=CHISQ.
```

Mann-Whitney *U* Test for Two Independent Samples (EX16b.SAV)

```
NPAR TESTS M-W=INCOME BY SCHOOL(1,2).
```

Kruskal-Wallis Test for Several Independent Samples (EX16c.SAV)

```
NPAR TESTS K-W=SCORES BY INSTRUCTION(1,3).
```

Wilcoxon Signed Ranks Test for Two Related Samples (EX16d.SAV)

```
NPAR TESTS WILCOXON=BEFORE WITH AFTER (PAIRED).
```

Friedman Test for Several Related Samples (EX16e.SAV)

```
NPAR TESTS FRIEDMAN=MORNING AFTERNOON EVENING.
```

References

Akaike, H. (1973), Information theory and an extension of the maximum likelihood principle, in Petrov, B.N. and Csaki F., Eds., *Proceedings of the 2nd International Symposium on Information Theory*, 267–281, Budapest: Akademiai Kiado.

Akaike, H. (1987), Factor analysis and AIC, *Psychometrika*, 52, 317–332.

Anderson, J.C. and Gerbing, D.W. (1988), Structural equation modeling in practice: a review and recommended two-step approach, *Psychological Bulletin*, 103, 411–423.

Arbuckle, J.L. and Wothke, W. (1999), *AMOS 4.0 User's Guide*, Chicago: SmallWaters Corporation.

Bentler, P.M. (1980), Comparative fit indexes in structural models, *Psychological Bulletin*, 107, 238–246.

Bentler, P.M. and Bonnett, D.G. (1980), Significant tests and goodness of fit in the analysis of covariance structures, *Psychological Bulletin*, 88, 588–606.

Bentler, P.M. and Mooijaart, A. (1989), Choice of structural model via parsimony: a rationale based on precision, *Psychological Bulletin*, 106, 315–317.

BMPD Statistical Software (1992), *BMPD Statistical Software Manual*, Release 7, Vol. 1 and Vol. 2, Los Angeles.

Browne, M.W. and Cudeck, R. (1993), Alternative ways of assessing model fit, in Bollen, K.A. and Long, J.S., Eds., *Testing Structural Equation Models*, Newbury Park, California: Sage, pp. 136–162.

Cattel, R.B. (1966), The scree test for the number of factors, *Multivariate Behavioral Research* (April 1966), 245–276.

Chou, C.-P. and Bentler, P.M. (1993), Invariant standardized estimated parameter change for model modification in covariance structural analysis, *Multivariate Behavioral Research*, 28, 97–110.

Cohen, J. and Cohen, P. (1983), *Applied Multiple Regression/Correlation Analysis for the Behavioral Sciences,* 2nd ed., Hillsdale, NJ: Lawrence Erlbaum.

Gerbing, D.W. and Anderson, J.C. (1984), On the meaning of within-factor correlated measurement errors, *Journal of Consumer Research*, 11, 572–580.

Hair, J.F., Anderson, R.E., Tatham, R.L., and Black, W.C. (1995), *Multivariate Data Analysis with Readings*, 4th ed., Englewood Cliffs, NJ: Prentice-Hall.

Ho, R. (1998), Assessing attitudes toward euthanasia: an analysis of the sub-categorical approach to right to die issues, *Personality and Individual Differences*, 25, 719–734.

Ho, R. (1989), Why do people smoke?: Motives for the maintenance of smoking behaviour and its possible cessation, *Australian Psychologist*, 24(3), 385-400.

Ho, R. and Venus, M. (1995), Reactions to a battered woman who kills her abusive spouse: an attributional analysis, *Australian Journal of Psychology*, 47(3), 153–159.

Ho, R. (1999), Factors influencing decisions to terminate life: condition of suffering and the identity of the terminally ill, *Australian Journal of Social Issues*, 34, 25–41.

Hoyle, R.H. and Smith, G.T. (1994), Formulating clinical research hypotheses as structural equation models: a conceptual overview, *Journal of Consulting and Clinical Psychology*, 62(3), 429–440.

Hu, L.T. and Bentler, P.M. (1998), Fit indices in covariance structure modeling: sensitivity to underparameterized model misspecification, *Psychological Methods*, 3, 424–453.

Hu, L.T. and Bentler, P.M. (1999), Cutoff criteria for fit indices in covariance structure analysis: conventional criteria versus new alternatives, *Structural Equation Modeling*, 6, 1–55.

Jöreskog, K.G. and Sörbom, D. (1989), *LISREL 8: A Guide to Program and Applications*, Chicago: SPSS.

Jöreskog, K.G. and Sörbom, D. (1993), *LISREL 8: Structural Equation Modeling with the SIMPLIS Command Language*, Hillsdale, NJ: Lawrence Erlbaum Associates.

MacCullum, R. (1986), Specification searches in covariance structure modeling, *Psychological Bulletin*, 100, 107–120.

MacCallum, R.C., Browne, M.W., and Sugawara, H.M. (1996), Power analysis and determination of sample size for covariance structure modeling, *Psychological Methods*, 1, 130–149.

Marsh, H.W., Hau, K-T., and Wen, Z. (2004), In search of golden rules: Comment on hypothesis-testing approaches to setting cutoff values for fit indexes and dangers in overgeneralizing Hu and Bentler's (1999) findings, *Structural Equation Modeling*, 11(3), 320–341.

McDonald, R.P. and Marsh, H.W. (1990), Choosing a multivariate model: noncentrality and goodness-of-fit, *Psychological Bulletin*, 107, 247–255.

Muthuen, B. and Kaplan, D. (1985), A comparison of methodologies for the factor analysis of nonnormal Likert variables, *British Journal of Mathematical and Statistical Psychology*, 38, 171–189.

Pedhazur, E.J. (1997), *Multiple Regression in Behavioral Research: Explanation and Prediction*, 3rd ed., New York: Harcourt Brace College.

Russell, D.W., Kahn, J.H., Spoth, R., and Altmaier, E.M. (1998), Analyzing data from experimental studies: a latent variable structural equation modelling approach, *Journal of Counseling Psychology*, 45(1), 18–29.

SPSS (1998), *Statistical Package for the Social Sciences*, Chicago.

Tabachnick, B.G. and Fidell, L.S. (1989), *Using Multivariate Statistics*, 2nd ed., New York: Harper Collins.

Wang, L.L., Fan, X., and Wilson, V.L. (1996), Effects of nonnormal data on parameter estimates for a model with latent and manifest variables: an empirical study, *Structural Equation Modeling*, 3(3), 228–247.

Williams, L.J. and Holahan, P.J. (1994), Parsimony-based fit indices for multiple-indicator models, *Structural Equation Modeling*, 1(2), 161–189.

Index

A

Absolute fit measures, 284
Aggression, 203, 282, 283
AIC, *see* Akaike Information Criterion
Akaike Information Criterion (AIC), 286, 317
AMOS graphics, 292, 293–296, 326
Analysis of variance (ANOVA), 51, 258, 282
Analysis of variance, factorial, 9, 57–84
    aim, 57
    assumptions, 58
    requirements, 57
    three-way factorial, 71–84
        data entry format, 72
        results and interpretation, 80–84
        SPSS output, 77–79
        SPSS syntax method, 77
        Windows method, 72–76
    two-way factorial, 58–71
        data entry format, 58
        data transformation, 65–70
        post hoc test for simple effects, 64
        results and interpretation, 63, 71
        SPSS output, 62–63, 71
        SPSS syntax method, 62
        Windows method, 59–61
Analysis of variance, one-way, 9, 51–56
    aim, 51
    assumptions, 51
    example, 52–56
        data entry format, 52
        post hoc comparisons, 56
        results and interpretation, 56
        SPSS output, 55
        SPSS syntax method, 54
        Windows method, 52–54
    requirements, 51
ANOVA, *see* Analysis of variance
Authoritarianism, 203, 282, 283

B

Backward deletion, statistical regression and, 246, 247
Bartlett's test of sphericity, 211, 215, 218, 232

Before and after design, 47
Beta coefficient, 201
Beta weights, 259, 267
Between-subjects effects, tests of, 78, 92, 104, 124, 151, 156
Bivariate Correlations window, 189, 191
Bonferroni type adjustment, 121, 131, 145, 160
Boolean factor analysis, 204

C

Cell
    chi-square, 365
    descriptive statistics, 90
    frequencies, 40
CFA, *see* Confirmatory factor analysis
CFI, *see* Comparative Fit Index
Characteristics, difference in, 4
Child care services, 23
Chi-square cell, 365
Chi-square goodness of fit, 9, 297, 312, 333
Chi-square statistic, 284, 367
Code book, 12, 13
Coefficient
    Display Format, 213, 225
    standard error of, 298
    table, 267, 278
Collinearity diagnostics, 252, 256, 269, 274
Communality of variable, 219
Comparative Fit Index (CFI), 285
Comparisons, pairwise, 137, 138
Component matrix, 230, 237
Compute Variable window, 65, 67, 87
Conceptual assumptions, factor analysis, 208
Condition index, 258
Confidence intervals, 131, 197, 262
Confirmation, subjecting hypotheses to, 3
Confirmatory factor analysis (CFA), 304, 321
Consistency, 239
Continuous variable, 16
Control group, 1
Correlated groups design, 47
Correlation, 183–194
    aim, 183–184